Computational Artificial Intelligence and Methods for industries

Vijayarangan Natarajan
Editor

Computational Artificial Intelligence and Methods for industries

A Machine-Generated Literature Overview

Springer

Editor
Vijayarangan Natarajan
IIT Madras Research Park
Tata Consultancy Services Limited (TCS)
Chennai, Tamil Nadu, India

ISBN 978-981-96-5276-1 ISBN 978-981-96-5277-8 (eBook)
https://doi.org/10.1007/978-981-96-5277-8

This Springer imprint is published by the registered company Springer Nature Singapore Pte Ltd.
The registered company address is: 152 Beach Road, #21-01/04 Gateway East, Singapore 189721, Singapore

Preface

The field of Computational Artificial Intelligence and Methods (CAMS) deals with artificial intelligence (AI) models for solving industrial challenges and improving the experience. This field involves multiple topics, such as tautology, reasoning, learning, algorithms, statistics, operation research, neural networks, problem solving, game theory, simulation, and queuing networks. It is not straightforward to observe, digest, analyze, and solve the industry problems for which we require CAMS along with the above topics as tools.

This book of Human–Machine collaboration describes various research papers on CAMS arising from different industries. It is a collection of research articles pertaining to CAMS covering fundamental, functional, applications, and challenges. There are four chapters presented in this book: Overview of CAMS, Computations and methods, Industrial challenges and solutions, and Future trends. Each chapter is written on the basis of theoretical and practical problems and solutions serving academia and industry.

This book has an interdisciplinary type of article, and the readers should have industry domain knowledge, fundamentals of artificial intelligence, information theory, computer networks, machine learning, and mathematics. This book is intended for graduate students, research scholars, academicians, and industry professionals.

The author of this book is grateful to God, family members, colleagues, friends, and publisher who encouraged a lot to investigate in a new era of CAMS with a fathom of learning. This book is dedicated to them.

Special thanks to Shrutika Dasari of Springer Nature for editing the entire manuscript.

Chennai, Tamil Nadu, India — Vijayarangan Natarajan

About This Book

Auto-summaries can be generated by either an abstractive or extractive auto-summarization:

- An extraction-based summarizer identifies the most important sentences of a text and uses the original sentences to create the summary.
- An abstraction-based summarizer creates new text based on deep learning. New phrases are created to summarize the content.

The auto-summaries you will find in this book have been generated via an extractive summarization approach.

Each chapter was carefully edited by Dr. Vijayarangan Natarajan. The editors selected the papers which were then auto-summarized. The editors have not edited the auto-summaries due to the extraction-based approach and have not changed the original sentences. You will find the editors' reviews and guidance on the auto-summaries in their chapter introductions.

In machine-generated books, editors are defined as those who curate the content for the book by selecting the papers to be auto-summarized and by organizing the output into a meaningful order. Next to the thoughtful curation of the papers, editors should guide the readers through the auto-summaries and make transparent why they selected the papers.

The ultimate goal is to provide a current literature review of Springer Nature publications on a given topic in order to support readers in overcoming information overload and to help them dive into a topic faster; to identify interdisciplinary overlaps; and to present papers which might not have been on the readers' radar.

Please note that the selected papers are not used to train a LLM while the auto-summaries are created.

Contents

About the Editor

Vijayarangan Natarajan obtained his Ph.D. (Mathematics) in 2001 at the Ramanujan Institute for Advanced Studies in Mathematics, University of Madras, Chennai. His doctoral research work was on Krein H*, J*-algebras, and triple systems. His research work includes mobile computing, Hilbert algebras, Jordan algebras, Lie algebras, number theory, elliptic curve cryptography, communication protocols, quantitative analysis, applications real and complex analysis in image processing, artificial intelligence, machine learning predictions, stochastic computing, etc. He has also received the Best Research Paper Award given by the Ramanujan Mathematical Society in June 2000. He has a rich flavor of academic and industry experience of approximately 30 years. He has presented his research work at several universities globally and published more than 100 articles, including papers, patents, and books. He is also fond of physical fitness.

Chapter 1
Overview of CAMS

Introduction by the Editor

Computational, artificial intelligence and machine learning systems (CAMS) are the latest trends in solving core research problems in cutting-edge industries. The prerequisite to study this chapter would require the fundamentals of Artificial Intelligence along with Mathematics stuff (Probability, Statistics, Algebra, Matrix, and Calculus). This chapter will showcase how we indulge research, innovate, and develop industrial products and solutions by means of advanced numerical methods, computational techniques, prediction, and monitoring. This chapter focuses on the trade space framework for aircraft manufacturing systems, deep neural networks for child's independent learning, data modeling in additive manufacturing, manufacturing of aerospace components through data models, a survey on the progress of artificial intelligence, a comprehensive study on the Internet of Things and biological computations. A subject Mathematical modeling like a reptile which moves anywhere else and survives at any conditions. Data modeling is an artificial intelligence reptile that applies the presence of mind to overcome CAMS. That is why we need to catch the intelligence reptile and observe its movements. This chapter provides an elegant way of designing systems for CAMSs.

CAMS is also to investigate to what extent AI models can interact with a designer to explore future forms of human–machine interaction, including autonomous and participative design. The study identifies and applies AI models to simulate three distinct learning mechanisms: design expertise, playfulness, and analogical reasoning. Each strategy has been applied to train different AI models, including generative models and reinforcement learning agents. CAMS helps to evaluate key performance indicators to achieve the optimal system performance. Of course, it is a highly complex process involving multidisciplinary stakeholders, various digital tools, and protocols. To address the complex issue, CAMS proposes a trade space framework based on semantic technology and Model-based systems engineering.

V. Natarajan (ed.), *Computational Artificial Intelligence and Methods for industries*, https://doi.org/10.1007/978-981-96-5277-8_1

CAMS proposes Convolutional Neural Networks (CNNs) for image contents extraction, whereas; Long Short-term Memory (LSTM) language model customizes to generate understandable and attractive text from Stanford Dogs Dataset (SDD) images. CAMS proposes an efficient way to translate the high-dimensional video image data into a lower dimensional format (the K-function curves), but also more effective than benchmark methods in detecting departures from a stable and in-control state by means of spatial data modeling via Ripley's K-function with Functional Analysis of Variance (FANOVA). The K-function performs to analyze the spatial pattern information for better understanding in the process behavior. The method is used in clusters or clouds of spatial domain. In this chapter, Additive Manufacturing via Laser Powder Bed Fusion (L-PBF) has been discussed. Recently, AI modeling on video images through Ripley's K-function on spatial domain is an advanced method or system that can be applied to synthesize and analyze spatter patterns in additive manufacturing. Further, this methodology leverages spatial statistics to understand the distribution and clustering of points, which in this context, refer to spatter particles. Ripley's K-function is a statistical function applied to analyze the spatial distribution of points. Ripley's K-function is a statistical function applied to analyze the spatial distribution of points. It determines the spatial points which are either randomly distributed or clustered, or regularly spaced. Ripley's K-function is a product of the inverse function of average density of the spatial points and the distance among the clustering of points. A Deep Neural Combinational Model (DNCM) has been applied to a digital image descriptor suited to improve pedagogy's learning model and DNCM provides recommendations and education innovations to children. The 3 important components of DNCM are CNNs, Combinational logic and Learning algorithms. The key features of DNCM are image processing, descriptor generation and personalization. which are explained deeply in this chapter. This conceptual framework to DNCM for digital image descriptors in child's independent learning can potentially revolutionize personalized education by means of the power of deep learning and combinational logic.

In this chapter, "An industrial portrait background removal solution based on knowledge infusion" provides a novel method for ID photo background removal. This method mixes or combines smart image processing and prior knowledge, mimicking human attention in manual segmentation. Besides, the method includes generating a convex hull around the body, applying a body map and grab cut for finer segmentation, and using heuristic post-processing for corrections. The solution outperforms state-of-the-art techniques and has been adopted industrially. Let us focus on leveraging the Sustainable Development Goals (SDGs) to bridge science, technology, and innovation. It uses machine learning (ML) to classify the SDG relevance in patent documents, employing patents registered in the European Patent Office (EPO) as a proxy for technological development. Further, the study examines how patents address SDGs, with a case study highlighting the ML model's efficacy in detecting SDG-oriented patents. The findings aim to enhance the alignment of R&D efforts, patent strategies, and the measurement and management of contributions to SDGs.

In response to global goals and sustainable development initiatives, the findings can advance the identification challenges of science and technology artifacts. Furthermore, CAMS offers input towards the alignment of R&D efforts and patenting strategies as well as measurement and management of their contribution to the realization of sustainable development goals (SDGs).

In this chapter, we outline the benefits of smart technologies in manufacturing, specifically focusing on the aerospace sector, which has been slower in adopting these technologies. It introduces a framework or methodology for assessing and performing the smart level of data-driven processes in aerospace manufacturing, which is not previously available in the literature. This framework drives companies or corporations to develop a roadmap for achieving higher levels of digitization by following an inductive rationale and using a case study to identify necessary features. The framework involves evaluating every process activity for Industry 4.0 properties and then analyzing the overall process. The AI data-driven approach enhances performance and interconnection levels. The framework was tested with both point-in-time and continuous timing, leading to significant improvements in digitally enhanced activities and monitored performance areas.

CAMS describes three different framings of computational thinking: cognitive, situated, and critical. Each framing offers valuable, but different insights of computational thinking. It is explained in this chapter. In addition to this, CAMS investigates how industrial equipment manufacturers can align the development of technology, business models and ecosystem relationships for the advancement of autonomous solutions.

This chapter presents gaps and future research opportunities concerning prediction and learning, decision making and optimization in view of intelligent decision making in today's era of uncertainty. In earlier days, we used Operations Research for Decision Support Systems (DSSs). Nowadays, CAMS helps DSSs in terms of time management, performance, and accuracy.

CAMS facilitates in designing Cross Reality (XR) system for content creation which can be automated for short-lived tasks and how seamless interoperability between devices facilitates skills transfer in human-machine hybrid environments.

This chapter describes Artificial intelligence (AI)-based rotor fault diagnosis (RFD) which poses a variety of challenges to the prognostics and health management (PHM) of the industry 4.0 revolution. CAMS introduces a vision-based method to scan, filter, segment, and correlate in real-time the physical printed part with the digital 3D model as well as to evaluate the performance of the additive manufacturing process. Further, high-resolution point cloud data of the printed part are automatically captured, filtered, segmented, reconstructed, and compared with the corresponding digital 3D models.

Machine generated Summaries

Disclaimer: The summaries in this chapter were generated from Springer Nature publications using extractive AI auto-summarization: An extraction-based summarizer aims to identify the most important sentences of a text using an algorithm and uses those original sentences to create the auto-summary (unlike generative AI). As the constituted sentences are machine selected, they may not fully reflect the body of the work, so we strongly advise that the original content is read and cited. The auto generated summaries were curated by the editor to meet Springer Nature publication standards. To cite this content, please refer to the original papers.

Machine generated keywords: manufacturing, business, image, design, automation, science, understand, assembly, computational, review, computer, industrial, theoretical, learn, manufacturing process

Expertise, Playfulness and Analogical Reasoning: Three Strategies to Train Artificial Intelligence for Design Applications [1]

This is a machine-generated summary of:

Mirra, Gabriele; Pugnale, Alberto: Expertise, playfulness and analogical reasoning: three strategies to train Artificial Intelligence for design applications [1]
Published in: Architecture, Structures and Construction (2022)
Link to original: https://doi.org/10.1007/s44150-022-00035-y

If you want to cite the papers, please refer to the original.

For technical reasons we could not place the page where the original quote is coming from.

Abstract-Summary
"The aim of this study is also to investigate to what extent AI models can interact with a designer to explore future forms of human–machine interaction, including autonomous and participative design."

"This study identifies and applies AI models to simulate three distinct learning mechanisms: design expertise, playfulness and analogical reasoning."

"Each strategy has been applied to train different AI models, including generative models and reinforcement learning agents."

"In the first application, the AI model extracts visual features from a dataset of shell and spatial structures, and then recombines such features to generate new design propositions."

"The third application illustrates that AI can be trained to discover meaningful features from biological forms and generate simple design objects through the visual abstraction of such forms."

"The applications demonstrate the ability of AI to synthesise design options and interact with a designer through visual data formats, such as 2D images and 3D models."

Introduction

"This research work presents a theoretical framework for implementing and applying AI models that interact with a designer to suggest design options."

"We use case studies to uncover the general properties of AI models that can be leveraged by CAD to support conceptual designs."

"Understanding how and what an AI model learns is important because, since the very first development of models for data generation [2], the design community has demonstrated a growing interest in these techniques."

"We believe that a better understanding of AI is crucial to exploit the full potential of these models for design applications."

"We hypothesise that AI models can support design applications at a deeper level."

"To test our hypothesis, we addressed the following three questions: which AI models are suitable for design applications?"

"The current research into AI for design applications does not seem to have been developed in continuity with this background."

The Different Meanings of AI in Design

"The narrative of this introduction is based on the evolution of the concept of 'intelligence', i.e. the many meanings intelligence has taken on over the years, and how such understandings have informed the development of AI applications in the design field."

"These I-CAD systems suffered from the same limitations as all early AI techniques: the categorisation of design knowledge was arbitrary and the query procedures were based on pre-defined rules."

"Independently of the design field in which ANNs have been applied, we have observed that the current research works are focused more on developing task-specific design tools than on artificial design assistants that actively participate in the design process."

"The aim of the early research in AI for design purposes was to produce functional CAD systems and develop models for future AI-supported creative design."

The Three Learning Mechanisms

"We expect that an AI model that is able to acquire design knowledge by simulating these mechanisms would build a form of 'design understanding' that is substantially different from that of human designers."

"The first learning mechanism – expertise – concerns the acquisition of knowledge from design precedents."

"Analogical reasoning is a learning mechanism that involves recognising similarities and transferring knowledge from a source domain to a target domain [3]."

"A notable example of analogical reasoning is the transfer of knowledge from biology to design, which is generally referred to as bio-informed design [4]."

"Some may be from other objects such as the clinker-built hulls of boats and so on", which suggests that expertise might involve transferring knowledge between two different design domains and – therefore – learning through analogical reasoning."

Applications

"The environment of an MDDP can be interpreted in many different ways: it can be (1) a conceptual space in which the designer performs design moves [5]; (2) a working environment in which the designer cooperates with other designers – i.e. with other agents – and/or negotiates with the client; or (3) a drawing board or spatial grid in which actions correspond to placing either strokes on paper or blocks in space to form a drawing or a spatial configuration."

"The implementation of the technique – named SPIRAL – involves combining reinforcement learning with generative models to train AI agents in image synthesis."

"We here describe our implementation of SPIRAL whereby the agent interacts with a 3D modelling environment – and therefore synthesises 3D models instead of images – to design artificial replacements of natural habitats."

Discussion

“With the work of the CAD and I-CAD research pioneers, we have developed a theoretical framework to implement AI models which has allowed us to set some key requirements in terms of design knowledge acquisition and communication.”

“It shows the potential of a variety of techniques and how such techniques could enable new forms of human-AI design partnership.”

“Our contribution in this area consisted in testing different forms of design exploration via simple AI-CAD interfaces.”

“In the first application, the designer could decide how many times a 2D footprint had to be interpreted by the AI model.”

“In the second application, the designer could ask the AI agent not to complete the design but to continue it for a limited number of steps.”

“These features enable a form of communication between designers and AI that is not deterministic but encourages exploration and revision of design ideas.”

Conclusion

“This study demonstrates that AI can simulate aspects of human cognition and support conceptual design and idea generation, by interacting with the designer through visual data formats, such as 2D images and 3D models.”

“We argue that it is important to understand how AI can be integrated with CAD systems to explore new and enhanced forms of human–machine interaction in design disciplines.”

"We here contribute to exploring emerging AI techniques in design applications and the integration of AI into CAD systems.”

“The aim of this work is not to demonstrate that AI supports design exploration but instead to show that different forms of exploration from those that characterise the current computational design tools and approaches can be used.”

“This work offers a theoretical framework, supported by practical applications, that can be used to implement AI models in architectural and structural design applications.”

A Semantic-Driven Tradespace Framework to Accelerate Aircraft Manufacturing System Design [6]

This is a machine-generated summary of:

Zheng, Xiaochen; Hu, Xiaodu; Arista, Rebeca; Lu, Jinzhi; Sorvari, Jyri; Lentes, Joachim; Ubis, Fernando; Kiritsis, Dimitris: A semantic-driven tradespace framework to accelerate aircraft manufacturing system design [6]

Published in: Journal of Intelligent Manufacturing (2022)
Link to original: https://doi.org/10.1007/s10845-022-02043-7

If you want to cite the papers, please refer to the original.
For technical reasons we could not place the page where the original quote is coming from.

Abstract-Summary
"During the design phase of an aircraft manufacturing system, different industrial scenarios need to be evaluated according to key performance indicators to achieve the optimal system performance."

"To address the digital discontinuity challenge during this process, this paper proposes a tradespace framework based on semantic technology and model-based systems engineering."

"It aims at functionality integration of requirement management, architecture definition, manufacturing system design, solution verification and visualization."

"An application ontology is developed to integrate assembly system domain knowledge, industrial requirements and system architecture model information."

"The proposed framework is implemented in a case study to support the fuselage orbital joint process design, which is part of the aircraft final assembly line."

"Different manufacturing system architectures are first designed by industrial system engineers supported by the application ontology stored in a graph database."

Introduction

"Model Based Systems Engineering (MBSE) uses models in a formalized way to support system requirements, design, analysis, verification and validation activities throughout all lifecycle phases of a system [7]."

"MBSE supports system development mainly by structural formalisms, while the domain specific knowledge of a complex system corresponding to the MBSE models is difficult to be captured and integrated [8]."

"There is a gap in the aircraft production field to integrate necessary functional components for conducting trade-offs during the manufacturing system design, from requirement definition to system design and simulations."

"Aiming at addressing the aforementioned challenges, this paper proposes a semantic-driven tradespace framework by converging ontology engineering and MBSE to optimize the digital continuity and interoperability in aircraft manufacturing system design, among which the ontology serves as a semantic core to integrate the modeling, simulation and requirement management activities."

Related Work

"Focusing on the conceptual design of aeronautical assembly lines, Mas and others [9] proposed a modelling approach to represent the process of aircraft assembly line designing based on the Airbus product lifecycle definition."

"To address this problem, it is necessary to develop integrated tools to support the assembly system design in collaboration with the product functional and industrial design [9, 10]."

"Model they developed an application ontology to support quality optimization of discrete manufacturing processes."

"Focusing on the aerospace production sector, in a recent study Arista and others [11] explored the feasibility of using industrial ontology to support aerospace assembly line design process."

"Mas and others [12] proposed an ontology-based Models-for-Manufacturing (MfM) methodology to support the conceptual design of aerospace assembly lines in Airbus."

Semantic-Driven Tradespace Framework

"The Requirement Management Module enables requirement definition and requirement tracing; the Architecture Definition Module aims for industrial system definition and trade-off scenario definition; the Verification Module contains different kinds of simulations to evaluate the generated solutions by reasoning in the Ontology layer; and the Visualization Module presents the simulation results and trade-off results according to different indicators to the end user for decision-makings."

"During the development of a new assembly system, the Requirement Management module will define the industrial requirements for the new system and provide relevant information to the ontology for creating a requirement model."

"The System Integration Module maps to the Model Management functional layer; the Requirement Management Module, Architecture Definition Module and

Verification Module are mapped to the Service Management layer; and the Visualization Module corresponds to the User Interaction layer of CDT."

Case Study

"The blue points on the left represent the different models that can be created as design options; the gray points represent the operations included in the model, among which the above two rows represent the two manual quarter-orbital level processes and the ones on the bottom correspond to the automatic half-orbital level process; the green points in the middle are the resources required by the operations; and the lines between points represent their relationships."

"All the points and lines have their own properties inherited from the application ontology, such as operation duration, resource cost, required resource quantity, etc. The data of the models generated by the engineer are stored in the Neo4j database and can be retrieved and exported in different formats, such as JSON, RDF and Turtle, etc. The assembly process design option can be retrieved and imported to the verification module for DES and 3D simulations."

Conclusion and Future Work

"It proves the feasibility of using an integrated application ontology to link relevant cross-domain information and knowledge, such as physical assembly system domain knowledge, industrial system requirements, architecture models information, etc. The ontology-based graph database can accelerate the generation of alternative design solutions for new assembly systems, which can be further verified through DES and 3D simulations."

"There are still multiple gaps to be bridged in future studies: Following the CDT reference architecture, the proposed tradespace framework can be extend to include more assembly stations and assembly lines; and to link future lifecycle phases aiming to create a complete cognitive digital twin of the assembly system."

"The GOPPRRE ontology is used in this study to represent the model information and linked with the domain knowledge following the IOF-Core and BFO principle."

"Similar approaches need to be generalized to create tool-agnostic methods for semantically represent the MBSE models."

Deep Neural Combinational Model (DNCM): Digital Image Descriptor for Child's Independent Learning [13]

This is a machine-generated summary of:

Naqvi, Nuzhat; Islam, M. Shujah; Iqbal, Mansoor; Kanwal, Shamsa; Khan, Asad; Ye, ZhongFu: Deep neural combinational model (DNCM): digital image descriptor for child's independent learning [13]

Published in: Multimedia Tools and Applications (2022)
Link to original: https://doi.org/10.1007/s11042-022-12291-7

If you want to cite the papers, please refer to the original.

For technical reasons we could not place the page where the original quote is coming from.

Abstract-Summary

"The art of preschool education to young children has become rare or vanished."

"Automatic image descriptors (captioning models) have recently shown their effectiveness, motivating us to utilize such models for address purposes."

"Developed image descriptors produce only complex and generic visual descriptions irrelevant to children's understanding."

"It is important to have a suitable image descriptor as teaching material for young children at the initial educational stage."

"To fill this gap, we introduced a novel digital image descriptor and 3k-Flickr-SDD dataset using smart augmentation that originally extracted and labeled solitary dogs' images from Flickr8k and Stanford Dogs Dataset (SDD) datasets."

"The proposed method accumulates Convolutional Neural Networks (CNNs) for image contents extraction, whereas; Long Short-term Memory (LSTM) language model customizes to generate understandable and attractive text from Dogs images."

Introduction

"This digital generation gap remains wide open as the older generation is unable to relevant learning material for preschool children that is available digitally."

"Young children are fascinated by digital images without knowing that available images are based on their understanding."

"It is common to perform transfer learning with predictive modeling problems that use image data as input [14, 15]."

"Image labeling has emerged as a challenging and important research area that generates a textual description of an image using computer vision (CV) and natural language processing (NLP) models."

"The sentence could be roughly generated through; retrieval, template, and deep convolutional neural-based methods."

"To resolve this twofold image captioning problem, we proposed an automatic digital image descriptor that primarily used smart augmentation and developed a merge 3k-Flickr-SDD dataset from Flickr8k and Stanford Dogs Dataset (SDD)."

Related Literature

"Kuznetsova and others [16] parsing-based language models have been introduced in image captioning, which is more powerful than static template-based methods."

"These methods can't generate variable length captions and are strictly constrained to image contents recognized by visual models."

"Retrieval-based methods Gong and others [17] first find visually similar images with captions from the training dataset."

"Chen and others [18] proposed an r attention-based image captioning method."

"The existing attention-based image captioning methods only consider spatial information for generating an attention map."

"Fu and others [19] proposed an alternative attention-based image captioning method."

"From all of the above-elaborated methods, we can observe that the process of image caption generation can be broadly divided into several parts, including visual feature extraction from the raw images, information fusion of visual features and attribute labels, and descriptive sentence generation."

Preliminaries

"It is not proficient for memorizing previous time steps by creating a loop within the network that allows information."

"Basic RNNs are networks of neuron-like nodes systematized into successive layers where each node in a given layer is linked with a one-way connection."

"Every other node in the next successive layer has a time-varying real-valued activation, and this neuron modifiable real-valued weight [20]."

"Unluckily, as that gap rises, RNNs develop impotent to learn to connect the information."

"This paper attempts to build a deep model that extracts visual information from images and converts it into general English sentences."

Proposed Model Architecture

"Feature extraction is performed from images whereas, the sequence processor and sentence generator employ for the language phase."

"The SDD holds visuals of 120 breeds of dogs, and this dataset [21] was constructed using images and annotation from ImageNet for the mission of fine-grained visuals classification, including 120 multiple categories."

"The 3k-Flickr-SDD dataset is a merged dataset containing only the dog's images extracted from two standard datasets called Flickr8k and SDD."

"A comprehensive version of the 3k-Flickr-SDD Dataset was developed for the generic use of the proposed model, which labels the unlabeled SDD dataset [22] and includes an attributive mechanism in each image description color, breed, and action information."

"These word sequences convert into sentences against images."

"Sequence processors utilize the LSTM network to predict each word within a sentence during training."

"The image and the words are mapped to the same space, the image using a vision CNN, and the words using word version LSTM."

Evaluation Analysis

"The criteria to judge the qualitative result of an image processing system is based on how many machine-generated sentences are correct in a similarity of reference sentences that are human-defined labels."

"According to BLEU and Meteor rules, consider the proposed model translation sentence correct; if they predict the correct information of an object, attributes, and possible relationship as defined in the dataset."

"Outstanding result, DNCM + Xceptionv3 and DNCM + VGG19 generated semantically correct sentences keeping the information about the principal object (dog), quantity (dogs), and their action (stands) precisely."

"DNCM + VGG19 generated semantically correct sentences that kept the information about an object, quantity, and action precisely for an inclusive result."

"Generated sentences shows that DNCM + VGG19 model generate correct whereas, DNCM + VGG16, DNCM + ResNet50 and DNCM + MobileNetv2 translated sentences are partially correct for a given image."

Results and Discussion

"We trained and tested our proposed model on GPU with 64 GM memory along with CPU Intel i9-9900k."

"We extracted dog images from the Flickr8k and SDD datasets using smart augmentation."

"The model preprocessed dataset images and descriptions to clean them and make them suitable for pre-trained models before the actual training process."

"It removes punctuation marks, short-length words, and numbers, etc. For training phase model expects an input photo signal vector of 4096 dimensional managed by a dense layer to produce a 256 component depiction of the image."

"DNCM has been trained using Flickr8k, a simplified version of 3k-Flickr-SDD, and a comprehensive version of the 3k-Flickr-SDD dataset separately."

"We saved the best three models for the representation of each dataset."

Conclusion and Future Work

"The first issue of existing datasets has a general or complex description of images."

"We extracted dog images from two standard datasets: Flickr8k and Stanford Dogs Dataset (SDD), to fill this scheme requirement and merge them into the 3k-Flickr-SDD dataset."

"The merged dataset consists of two versions; a simplified version of the 3k-Flickr-SDD dataset eradicates the complexity from candidates' captions."

"The comprehensive version of the 3k-Flickr-SDD dataset emanates an attributive mechanism (such as breed, color, and action) for universal candidates' sentence generation."

Modeling Spatial Point Processes in Video-Imaging via Ripley's K-Function: An Application to Spatter Analysis in Additive Manufacturing [23]

This is a machine-generated summary of:

Colosimo, Bianca Maria; Pagani, Luca; Grasso, Marco: Modeling spatial point processes in video-imaging via Ripley's K-function: an application to spatter analysis in additive manufacturing [23]
Published in: Journal of Intelligent Manufacturing (2022)
Link to original: https://doi.org/10.1007/s10845-022-02055-3

If you want to cite the papers, please refer to the original.

For technical reasons we could not place the page where the original quote is coming from.

Abstract-Summary

"For an increasing number of applications, the quality and the stability of manufacturing processes can be determined via image and video-image data analysis and new techniques are required to extract and synthesize the relevant information content enclosed in big sensor data to draw conclusions about the process and the final part quality."

"This paper focuses on video image data where the phenomena under study is captured by a point process whose spatial signature is of interest."

"A novel approach is proposed which combines spatial data modeling via Ripley's K-function with Functional Analysis of Variance (FANOVA), i.e., Analysis of Variance on Functional data."

"The K-function allows to synthesize the spatial pattern information in a function while preserving the capability to capture changes in the process behavior."

Introduction

"Many authors outlined the need for appropriate statistical modeling and monitoring methods for image data [24–27], but still limited attention has been devoted to video-image data, where the underlying evolution of the process in space and along time has to be properly captured and monitored."

"Despite a large literature on K-functions and SPP methods for image data, there is a lack of applications involving video-image data, although the dimensionality reduction enabled by K-functional representations makes such approach particularly attractive in video-image processing and analysis too."

"This paper presents a method which combines an image segmentation and processing step to transform video-image data of spatter ejections into Ripley's K-function curves, and a curve fitting approach based on Functional Data Analysis [28], which enables the classification of different process behaviors via FANOVA."

Motivating Example

"The motivating example for the proposed approach regards the analysis of spatter ejections in L-PBF as a signature of the process behavior."

"Spatters, as many other relevant quantities, can be measured in-line and in-situ thanks to the layerwise paradigm that allows to observe (e.g., with high-speed cameras or thermal cameras) many process-related phenomena during the production of each layer [29–33]."

"An increasing number of studies have been devoted in the last years to the possibility of using spatters and other L-PBF by-products (i.e., vapor gas and plasma emissions) for in-line and in-situ process monitoring [34–38]."

"The mainstream analytic approach consists of estimating, frame by frame, various synthetic descriptors, e.g., the number of ejected spatters, their distance from the melt pool, their size, velocity, etc., and relating them to the variation of process conditions."

Proposed Methodology

"K-functions have different properties that make them suitable to capture different patterns of spatial point processes."

"The value of t at which the estimated K-function reaches the asymptote reflects the dispersion of the points, whereas the slope and changes of the first and second derivatives reflect whether the points are clustered together and their relative distances at different scales."

"The proposed approach involves an image pre-processing phase followed by the frame-by-frame estimation of the corresponding K-function."

"The operation is then repeated for all video frames, as a parametric model of estimated functions can be fitted to enable statistical inference of K-functional patterns."

"The segmented centers of mass were then used to estimate the Ripley's K-function corresponding to a specific frame."

"It is also possible to compare the K-function representation of spatter patterns in any new layer or any new build against a reference pattern, representative of a stable and in-control process state."

Results

"The analysis discussed in the following demonstrates the suitability of the proposed approach to this aim, where the spatter behavior observed with a volumetric energy density of 80 kJ/cm was used as a signature of reference process performance."

"A further analysis consists of determining the differences between each energy density level and a reference condition, i.e., the condition that produced the lowest porosity in the manufactured specimens."

"They fail in detecting a statistical difference in the spatter behavior between overmelting conditions (energy density levels 4–6) and the reference level 3."

"The adoption of the K-function to characterize the spatial spread of spatters, instead, highlighted a statistically significant difference between the most severe over-melting state and the reference, which was also confirmed by a higher porosity in the part."

Conclusions and Future Developments

"The application of the proposed approach in L-PBF showed that the data synthesis based on the K-functions representing the spatial spread of spatters allowed distinguishing between different process states that led to volumetric defects in the final part."

"The major strength points of the proposed methodology can be synthesized as follows: (i) enhanced in-situ detection of unstable process conditions: spatters are known to be a relevant signature for in-situ monitoring of the L-PBF process, and the proposed approach enables a more effective detection of departures from a stable process state than competing methods adopted in the mainstream literature; (ii) enhanced in-line data modeling for efficient process optimization, by replacing post-process inspection with in-line measurements, (iii) reduction of data handling and storage needs thanks to the massive data reduction enabled by the K-function modeling, moving from high-speed/high-resolution video-image streams to 1D functional representation of salient information."

An Industrial Portrait Background Removal Solution Based on Knowledge Infusion [39]

This is a machine-generated summary of:

Riad, Rabia; Ros, Frédéric; hajji, Mohamed El; Harba, Rachid: An industrial portrait background removal solution based on knowledge infusion [39]

Published in: Applied Intelligence (2022)
Link to original: https://doi.org/10.1007/s10489-021-03099-3

If you want to cite the papers, please refer to the original.
For technical reasons we could not place the page where the original quote is coming from.

Abstract-Summary
"Background removal of an identity (ID) picture consists in separating the foreground (face, body, hair and clothes) from the background of the image."

"State of the art image processing techniques encountered several segmentation issues and offer only partial solutions."

"Knowledge is infused by considering the spatial relation between anatomic elements of the ID image (face feature, forehead, body and hair) as well as their "signal properties"."

"The process consists in first determining a convex hull around the person's body including all the foreground while keeping very close to the contour between the background and the foreground."

"A body map generated from biometric analysis associated to an automatic grab cut process is applied to reach a finer segmentation."

"A heuristic-based post-processing step consisting in correcting potential hair and fine boundary issues leads to the final segmentation."

"Experimental results show that the newly proposed architecture achieves better performances than tested current state-of-the-art methodologies including active contours, generalist popular deep learning techniques, and also two other ones considered as the smartest for portrait segmentation."

Introduction

"Over and under segmentation are related to images having one of the following features: Insufficient, non uniform light, shadow, light reflection Blurred edge between the body and background Colored and textured hair with irregular shape and appearance, hair leaks Clothes with any appearance (color, texture...) With the evolution of deep learning (DL) techniques, a novel generation of image segmentation has been developed during the last few years."

"Innovations are proposed for segmenting an ID image at the pixel level into two objects; face, hair and clothes as one object representing the foreground and its surrounding as the background."

"The method is then based on a knowledge-infused approach that cleverly hybrids smart image processing tasks and this prior knowledge including the spatial relation between anatomic elements of the ID image (face feature, forehead, body and hair) as well as their "signal properties"."

Related Work

"Various algorithms have been proposed in the literature for image segmentation [40]."

"The most commonly used techniques are: automatic threshold segmentation such as Otsu's method, edge detection based segmentation such as active contour models (or snakes), graph-based segmentation techniques such as Lazy snapping and segmentation techniques based on clustering such as K-means."

"It is based on the Min-Cut/Max-Flow algorithm and has been extended to segment color images [41]."

"Many deep learning based image segmentation approaches often achieve the highest accuracy rates on popular datasets [42–44]."

"When it comes to segmentation of ID images, none of these techniques can conceptually deal with fine features and boundary details without a very specific architecture and/or a huge training set."

"In synthesis, this problem is known as a very difficult segmentation task where classical techniques face recurrent issues while several emerging techniques based on deep learning appear to be promising."

The Proposed Background Removal Method

"The process consists in determining a convex hull around the person's body that guarantees the inclusion of all the foreground elements while possibly includes some background pixels."

"A map of the ID image is created in the body map in order to set up the grab cut algorithm and apply it automatically."

"This last step allows to remove the remaining background pixels that cannot be discriminated by the convex hull neither by the grab cut algorithm."

"The Grabcut algorithm is applied using the input identity photo and its map in order to process a mask where the white (255) pixels correspond to the foreground region."

"For hair, there are two sub-series of seeds: (i) the points located beyond the face ellipse but inside a certain range that guarantees that all these seed points are not labeled as background pixels; (ii) The points between the edge of the current mask and the face ellipse."

Experimental Study and Results

"The evaluation techniques used in this work are the following: The Sum of Absolute Differences (SAD) is a simple measure of similarity between two images."

"Background removal is performed on the industrial database using four classical image segmentation techniques: Otsu's [45] method, K-means a clustering-based segmentation, Lazy snapping [46] and Active contour [47]."

"The first technique called U-Net, was developed by Ronneberger and others [48] for biomedical image segmentation."

"The two best state of the art techniques dedicated to portrait segmentation reach higher score than the ones of deep learning techniques."

"The proposed approach yielded a significantly better success rate than classical image segmentation techniques and all the popular deep learning Neural Networks."

"The low performances of generic deep learning techniques can be explained by the imbalance in the training set between common "cases" where ID images are relatively easy to segment and more complex "cases" that have high variability while being not enough represented."

Conclusion

"Experimental results show that the newly proposed architecture achieves a significantly better classification performance than all the current state-of-the-art methodologies including popular deep learning NNs."

"The promise of ANNs to automatically discover and extract useful feature patterns from data without dwelling on domain expertise is highly exciting."

"They ignore a very important modality, i.e. a priori domain expertise, and are unable to harvest any benefit from the context knowledge."

"They are firstly the key questions related to the domain knowledge itself, the difficulty in the representation of domain knowledge, the overlap of the domain knowledge and the knowledge that can easily be learned by deep learning."

"The idea consists in studying a novel approach that could incorporate expert knowledge into a deep NN (or several specialized sub NNs) with the objective of consolidating classification performances and reaching a partially explainable model."

Mapping the Sustainable Development Goals (SDGs) in Science, Technology and Innovation: Application of Machine Learning in SDG-Oriented Artefact Detection [49]

This is a machine-generated summary of:

Hajikhani, Arash; Suominen, Arho: Mapping the sustainable development goals (SDGs) in science, technology and innovation: application of machine learning in SDG-oriented artefact detection [49]
Published in: Scientometrics (2022)
Link to original: https://doi.org/10.1007/s11192-022-04358-x

If you want to cite the papers, please refer to the original.
For technical reasons we could not place the page where the original quote is coming from.

Abstract-Summary
"We utilize SDG classification of scientific publications to compile a machine learning (ML) model to classify the SDG relevancy in patent documents, used as a proxy of technology development."

"The ML model was used to classify a sample of patent families registered in the European Patent Office (EPO)."

"The analysis revealed the extent to which SDGs were addressed in patents."

"We also performed a case study to identify the offered extension of ML model detection regarding the SDG orientation of patents."

Introduction

"The post-2015 sustainable development agenda calls on all countries to enhance research, upgrade technological capabilities, encourage innovation and increase public and private investment [50]."

"Fostering innovation is part of Goal 9 related to resilient infrastructure and inclusive, sustainable industrialization, while Target 9.5 elevates the role of research and innovation policy well beyond STI as one of the means of implementation."

"In the United Nations general assembly briefing materials, the importance of science, technology and innovation (STI) for the SDGs has been numerously mentioned in yearly forums."

"A direct quote from Marie Chatardová, President of the Economic and Social Council, at the 2018 New York STI Forum "No one can ignore the vital role of science, technology and innovation (STI) in advancing the transformative impact"."

"This research focuses on analysing science and technology literature contribution to SDGs using machine learning methods."

Background

"The approach used is based on utilizing the classification of scientific documents based on their relevance to SDGs."

"While patent documents differ from scientific documents, previous research has shown that using natural language processing and machine learning can model the interaction between science and technology documents [51]."

"The approach used relies on an existing classification of scientific publications' relevance to SDGs as a gold standard, trains a model with scientific publications and transfers the model to patent data to create a classification of patent relevance."

"In the work, Yau and others [52] tested several approaches of topic modeling to scientific publication clustering and show the potential of methods such as LDA in producing meaningful research clusters."

"Yoon and Phaal [53] used a K-means approach to classify patent documents in the context of technology road mapping."

Methodology

"This research has benefited from the development of machine learning classification algorithms on texts to facilitate the construction of the SDG detecting model."

"We considered Naive Bayes Classifier for Multinomial Models, Linear Support Vector Machine, Logistic Regression as classifiers and TF-IDF, Word2vec and Doc2vec as text modeling strategies."

"Text vectorization techniques, namely TF-IDF, Bag of Words and vectorization, are very popular choices for machine learning algorithms and can help convert text to numeric feature vectors."

"With the feature implementation explained here, we will perform the classifier task with Naive Bayes Classifier for Multinomial Models, Linear Support Vector Machine and Logistic regression."

"Word2vec uses a large corpus of text and produces a vector space model where words that share common contexts in the corpus (i.e., being nearby words) are positioned close to each other [54]."

"Naive Bayes is one of the algorithms used in our research to train a machine learning classification model."

Analysis and Results

"Applying the ML model for detecting the SDG 7 categories in 132 k patent sample results in almost 10 thousand patent families, which have the probability of above 85% for SDG 7."

"Cross-referencing (1) SDG7 query-driven records, (2) ML model-driven records for SDG7 at a probability level 85% or above, we used the International Patent Classification (IPC) to check the number of patent families included in each patent class."

"This analysis suggests that both lexical query resulted patent families and ML decided patent families for SDG7 overlap overall, but there is some difference of coverage in terms of IPC classes as well."

"The ML model expanded the scope of IPC classes by 14%, where 75 new classes were identified to be relevant patent families for SDG7."

"The colour of the new keyword clusters identified by the ML model is marked as green while the overlapping clusters within two modes of SDG 7 identification patent families remained blue."

Discussion and Conclusion

"Based on the publications with the highest relatedness to SDGs, the study trains and creates a machine learning model to detect the relatedness of scientific publications and, finally, uses the approach to analyse patent documents."

"In order to test the validity of the ML model in detecting SDG contribution in patents, the classification was compared to the already existing lexical queries using patent classifications."

"Not all scientific and technological artefacts share the same features, and for example, patent documents, due to their technical nature, do not directly correspond to an SDG definition."

"While the linkage between science and technology and publication and patents is evident, we could utilize this connection and expand the identification of SDG oriented artefacts from publications to patents."

"We hope these findings make sound automated text classification more approachable to STI researchers and encourage future research to integrate novel approaches for complex classification tasks such as SDGs."

Assessment of Smart Transformation in the Manufacturing Process of Aerospace Components Through a Data-Driven Approach [55]

This is a machine-generated summary of:

Bernabei, Margherita; Eugeni, Marco; Gaudenzi, Paolo; Costantino, Francesco: Assessment of Smart Transformation in the Manufacturing Process of Aerospace Components Through a Data-Driven Approach [55]

Published in: Global Journal of Flexible Systems Management (2022)
Link to original: https://doi.org/10.1007/s40171-022-00328-7

If you want to cite the papers, please refer to the original.

For technical reasons we could not place the page where the original quote is coming from.

Abstract-Summary

"In such a sector, data-driven digitization initiatives play a significant role as they allow for flexibility, efficiently considering time and resources."

"The paper presents a framework to assess the smart level of data-driven processes in the aerospace sector since it is not available in the literature."

"The framework helps companies to draw a roadmap to achieve greater levels of digitization."

"An aerospace case study formalizes evidence and needs, identifying features to design the framework."

"It is based on 7 assessment steps, differentiated into two levels of detail of the process."

"The data-driven approach provides added values in terms of performance and level of interconnection ."

"The assessment highlights 10 out of 11 digitally enhanced activities, and the company moved from 5 to 22 monitored performance areas."

Introduction

"Data-driven approaches allow for flexibility, both considering organizational and technological aspects."

"The substantial use of data and information technologies enhances business process flexibility, which also impacts business alignment, communication, and collaboration [56]."

"Even if assessment tools exist to support 4.0 evolution, there are no frameworks designed to specifically support the aerospace sector, considering the implementation of data-driven 4.0 solutions."

"To address these gaps, the current paper contributes to research by providing a framework for the implementation of data-driven approaches in the aerospace industry and gathering insights from its application."

"The framework helped in drawing a roadmap for improvements and for company growth, also enabling a periodic analysis of the process to understand the data-driven smartness enhancement and to compare it with the expected level."

Literature Review

"With this purpose, a literature review was conducted to capture the current digital level of the aerospace sector, the benefits and challenges related to data-driven approaches, and the existing tools to support smart transformation."

"Today in the aerospace industry, there are problems associated with product quality, which can be improved by digital technologies [57]."

"IoT technologies are a potential solution to building a secure, reliable system architecture that gathers, organizes, and processes large amounts of data from disparate assets in the aerospace industry [58]."

"Considering advances in technology, computing capabilities, and improved access to large data sets, the data-driven approach increases."

"The aerospace sector needs extensive 4.0 solutions and data-driven ones are promising."

"The ability to implement data-driven approaches is enabled by advances in technology, computing capabilities, and access to large data sets."

Research Motivation and Innovativeness

"The specific nature of the framework, as well as the innovative aspects compared to tools in the literature, are summarized as follows: Assessment object: the MARS framework assesses the smart level improvements in implementing a specific 4.0 technology; Context of assessment: the focus is on one manufacturing process, not on the company or interconnected processes producing services and products; Timing of the assessment: the manufacturing process is assessed before the implementation of the 4.0 solution to get an "as is" situation and a "to be" projection after the implementation of the specific 4.0 technology, with a chance to repeat the assessment over time; Assessment modality: a qualitative assessment of the production process, either from an aggregate point of view (i.e., the overall process is considered) or from a disaggregated point of view (i.e., the process is decomposed into its activities and it is assessed accordingly)."

Research Approach: Context, Conceptualization, and Operationalization

"The needs are: Perform the first assessment of the process, before the introduction of the data-driven technologies, to understand the current smart level of the process; subsequently, re-assess the process considering the changes that new technologies introduced."

"Decline part of the evaluation on the individual process activities, especially those evaluations for which specific and detailed data-driven information is desired; According to more high-level and comprehensive criteria, evaluate the process as a whole, aggregating resources to provide an overall view."

"Meetings with the company revealed the need to develop a qualitative smart assessment, since the objective the company wanted to pursue was to evaluate and monitor strategic and technological factors internally, keep track of changes and improvements in the production process, checking for alignment with the defined strategy."

Results of the MARS Framework Applications

"In the second meeting, the company shared the results and justified the choices, showing coherence between what was declared, the current and the digital improves smart level of the manufacturing process."

"In internal and sustainability performances, the element at level 2 will be evaluated: Flexibility, Process, and product quality, Service, Productivity/production efficiency, Production capacity, Logistics, Maintenance, Manufacturing process reliability, Security and Safety, Social sustainability, Economic sustainability, and environmental sustainability."

"The current version of the process enabled the measurement of flexibility, quality, productivity, and capacity performances."

"The current interconnection level of the production process is represented by product and station."

"The company found smartness shortcomings of the production process and it evaluated the introduction of a new architecture for data collection."

"The framework results in a meaningful tool to support the evolution of data-based digitization, to guide in the assessment of the production process, and to provide practical insights."

Discussion and Evidence from the MARS Framework Application

"The MARS application and the results presented, provided the company insights on the level of smartness growth resulting from the implementation of a data-driven approach, i.e., a CPS-based data collection architecture."

"Looking at the single constituent activities of the process, the overall level moves from computerization, where data collection for specific activities is total and mainly digital, to visibility, where the collected data for specific activities are systematically analyzed according to specific purposes, and results are always available, i.e., they can be always accessed."

"By analyzing the results the company got confirmation that not all process activities benefit equally from innovation."

"The framework application showed the company that the CPS architecture allows the measurement of external performances (i.e., volume flexibility, reliability, maintainability, readiness, punctuality, regularity, and availability) that in the company's opinion could result in a more flexible process and in a greater focus on clients' needs."

Conclusions and Future Developments

"The paper contributes to research aimed at supporting the smart transformation of the aerospace industry, providing a framework for the implementation of a data-driven approach."

"The designed framework can be autonomously implemented, since it's intuitive and ready to use, providing a simple yet highly effective approach to get insights about the improvements resulting from a data-driven approach implementation in a company."

"The results show that the framework capture different aspects of the process, such as the difference in smart growth of the activities, but also transversal and high-level aspects."

"The framework design, following an inductive rationale, required the identification of a specific sector and production process."

"The framework can be applied to other industries for which a data-driven approach is desirable or required."

"The limitations of the proposed framework consist in its being designed to assess a well-defined manufacturing production process."

Survey: Artificial Intelligence, Computational Thinking and Learning [59]

This is a machine-generated summary of:

Dohn, Nina Bonderup; Kafai, Yasmin; Mørch, Anders; Ragni, Marco: Survey: Artificial Intelligence, Computational Thinking and Learning [59]

Published in: KI - Künstliche Intelligenz (2022)

Link to original: https://doi.org/10.1007/s13218-021-00751-5

If you want to cite the papers, please refer to the original.

For technical reasons we could not place the page where the original quote is coming from.

Abstract-Summary

"With the growing relevance of computational thinking, these two efforts have become more closely connected."

"We describe three different framings of computational thinking: cognitive, situated, and critical."

"Each framing offers valuable, but different insights into what computational thinking can and should be."

"We combine the three framings into one framework which emphasizes that (1) computational thinking activities involve engagement with algorithmic processes, and (2) the mere use of a digital artifact for an activity is not sufficient to count as computational thinking."

"We further present a set of approaches to learning computational thinking."

"We argue for the significance of computational thinking as regards artificial intelligence on three counts: (i) Human developers use computational thinking to create and develop artificial intelligence systems, (ii) understanding how humans learn can enrich artificial intelligence systems, and (iii) such enriched systems will be explainable."

The Significance of Learning Computational Thinking for the Field of AI

"While there are several definitions of AI, probably the most prominent one considers two dimensions, the first one focused on the shift from thinking to action, and the second one on the shift from human to rational agent [60]."

"Her much-cited definition of CT came in a follow-up article: "Computational Thinking is the thought processes involved in formulating problems and their solutions so that the solutions are represented in a form that can be effectively carried out by an information-processing agent... human or machine, or... a combination of humans and machines" [61]."

"We prefer a broader definition proposed by Dohn [62], our translation], which targets development of digital artifacts (including AI) more generally: "Computational thinking consists in the cognitive processes involved in the development of digital artifacts and programs to live in today's world.""

"Understanding how humans think computationally can provide new explainable models [63] of machine learning and thus push AI forward."

Different Framings of Computational Thinking

"In its original definition, CT was mostly framed in cognitive terms, as the connotation of 'thinking' in relation to 'computation' illustrates."

"The cognitive framing harks back to Wing's [64] early description of CT as "taking an approach to solving problems, designing systems and understanding human behaviour that draws on concepts fundamental to computing" (p. 33)."

"CT from a cognitive perspective [65] presents an individualistic framing of human and machine learning that downplays the social and cultural contexts in which it operates."

"A different framing of CT is computational participation [66, 67] highlighting learning's situated nature."

"Framings of CT as participation emphasize that the design of any system is situated within a social context that provides interaction, feedback on interactions and performances."

"Such framings draw on Hutchins' seminal work on cognition in the wild [68] that examined how culture shaped individual cognition and computational systems and saw them as part of a larger social context."

Delimitation of Computational Thinking Activities

"We discuss three key questions: A. Do CT activities (primarily) concern construction of or critical reflection on algorithmic artifacts?"

"The answer to question B is that CT activities need not involve digital artifacts because the algorithmic solution can be found and represented with analogue means (e.g., in the learner's mind or with pen and paper) which essentially will involve the same cognitive processes as the ones necessary for designing a programmed solution [69, 70]."

"The answer to question B is that CT activities necessarily involve digital artifacts."

"The critical framing is portrayed in the brace, spanning both the solid and dashed line figures, to indicate that critical reflection should concern not only the algorithms involved in CT activities but also the algorithms underlying pure use of digital artifacts."

Practical, Pedagogical Approaches and Applications for Learning Computational Thinking

"It encompasses a set of pedagogical takes which converge in emphasizing the following characteristics of learning processes as decisive in supporting the learning of CT: creativity, playfulness, learners' construction of digital artifacts, self-chosen projects, participation in construction practices, collaboration with others, and curation of learners' passion for their projects—what Resnick has termed the 4 p's of constructionism: Projects, peers, passion, and play [71]."

"New empirical questions can be raised about students' CT practices and learning of CT when the design environments provide automated scaffolding utilizing intelligent systems like recommender systems: Such a system can for instance inform the learners about what CT concepts they successfully demonstrate or alternatively fail to achieve when creating or modifying a game, for example following the approach of critiquing systems in domain-oriented design environments [72]."

AI Perspectives on Computational Thinking

"We focus on three points: (i) Human developers use CT to create and develop AI systems (correlating with (b)), (ii) Understanding how humans learn can enrich AI systems, (iii) Such enriched systems will be explainable, i.e., they will be able to explain their underlying reasoning (e.g., algorithms) to the end users."

"For both the AI frame problem and the common-sense problem, coming to understand how we humans develop CT skills across our lifetime may be the solution: It can signpost the way to develop intelligent systems that learn to solve problems like we do."

"To understand how humans learn to think computationally will provide AI with an approach to higher-level problems that may require more of an understanding than current systems provide."

"If we come to understand how humans do CT, we can enrich and augment state of the art AI towards systems that can perform symbolic machine learning—even with the aim of learning the skills humans apply to develop AI systems."

Overview of the Special Issue

"A common denominator of these articles is that they originate in an existing pedagogical approach to learning CT but bring a different theoretical perspective to its study."

"All three articles challenge the dominant cognitive conceptualization of this approach by applying a situated lens (corresponding more to the situated framing) and investigating problem-solving in its sociocultural and subject-didactical contexts."

"The fourth article, by Andreas Lindenskov Tamborg, Raimundo Elicer, and Daniel Spikol, investigates the relationship between mathematics and CT, and particularly how the integration of these disciplines can support students in AI learning."

"The fifth article, by Andrea Valente and Emanuela Marchetti presents a new conceptualization—along with its pedagogical implications—of the approach of iterative problem-solving with IT."

"The connection between CT, AI and pedagogical practice is highlighted in the articles by Tamborg and colleagues and by Valente and Marchetti in the following ways."

A Maturity Framework for Autonomous Solutions in Manufacturing Firms: The Interplay of Technology, Ecosystem, and Business Model [73]

This is a machine-generated summary of:

Thomson, Linus; Kamalaldin, Anmar; Sjödin, David; Parida, Vinit: A maturity framework for autonomous solutions in manufacturing firms: The interplay of technology, ecosystem, and business model [73]

Published in: International Entrepreneurship and Management Journal (2021)
Link to original: https://doi.org/10.1007/s11365-020-00717-3

If you want to cite the papers, please refer to the original.
For technical reasons we could not place the page where the original quote is coming from.

Abstract-Summary
"Significant advancements within the fields of digitalization, electrification, and automation have enabled the development, testing, and implementation of increasingly advanced autonomous solutions."

"Implemented autonomous solutions have rarely advanced beyond 'islands of autonomy'."

"There is a need to develop a business-focused maturity framework for autonomous solutions to contribute to a predominantly technical discourse and support equipment actors and their wider ecosystems in commercializing autonomous solutions."

"The purpose of this paper is to investigate how industrial equipment manufacturers can align the development of technology, business models and ecosystem relationships for the advancement of autonomous solutions."

"We capture our findings in a three-level maturity framework for industrial autonomous solutions."

"This framework unwraps the attributes of each level from the perspectives of technical system development, ecosystem configuration, and business model design and is complemented by three overarching principles for the successful commercialization of autonomous solutions."

Introduction

"It is not clear how the ecosystem configuration advances with the level of autonomous solution maturity, nor how the ecosystem is more broadly related to technology development."

"There is a need for a maturity framework to advance understanding of how firms can successfully develop and commercialize autonomous solutions, taking into consideration the importance of alignment between technological, ecosystem, and business model perspectives."

"To address these gaps, we developed the following research question: How can industrial equipment manufacturers align the development of technology, business models and ecosystem relationships for the advancement of autonomous solutions?"

"Our findings show that key activities are split across the three dimensions of technology development, ecosystem configuration, and business model design that evolve over three maturity levels of autonomous solutions (level 1: operator assistance, level 2: semi-autonomous operation, level 3: fully autonomous operation)."

Theoretical Background

"Autonomous solutions are an example of smart, connected products that are radically changing the roles of traditional manufacturers and redefining industry boundaries [74]."

"Despite the central role of technology, innovating autonomous solutions requires the focus to be shifted so that ecosystem and business model features are also considered."

"The development of autonomous solutions challenges equipment manufacturers to work across firm boundaries and consider a wider ecosystem of actors, systems, and products [75]."

"To truly leverage the benefits of increased connectivity and enable a system-of-systems approach, equipment manufacturers must seek to strengthen their autonomous solutions through connectivity to complementary system capabilities from other ecosystem actors "beyond firm boundaries and across networks in the form of collaborative value creation" [75], including possible collaboration with competitors [76]."

"There is a specific need to investigate appropriate configurations of autonomous solutions at different maturity levels in order to understand the commercialization challenges that industrial actors face."

Methods

"This leads us directly to the research question: How can industrial equipment manufacturers align the development of technology, business models and ecosystem relationships for the advancement of autonomous solutions?"

"The cases for this study were selected by theoretical sampling [77–79], and the selection criteria were informed by the study's purpose (which was to develop a maturity framework for industrial autonomous solutions, focused on the perspective of industrial equipment manufacturers)."

"The study's purpose captures three key aspects: the perspective of industrial equipment manufacturers, the context of autonomous-solutions provision, and the focus on maturity-level development of these solutions."

"We selected equipment manufacturers that are either developing, testing, and/or implementing a range of autonomous solutions at different autonomous maturity levels."

"This was facilitated by an interview guide, which was based around overarching themes such as the development and implementation of autonomous solutions, data and technology, ecosystem collaboration and competition, and business model design."

Findings: Detailing a Maturity Framework for Autonomous Solutions

"An automation customer confirmed the importance of operator-assisted functionalities and stressed that their continued development remains a key focus for achieving higher levels of autonomous equipment maturity, specifically referencing collision-awareness and collision-avoidance systems."

"The Chief Technological Officer from an SME supplier commented that development of the assistive solution will be tailored to meet the requirements of the equipment manufacturer, enhancing the value proposition for the end customer of the equipment through improved operator-assist functionalities."

"Connectivity standards were developed from close collaboration between the equipment manufacturer and the connectivity supplier, resulting in the end customer receiving a networked solution tailored to the requirements of its autonomous equipment operations."

"Consideration of these overarching principles can assist equipment manufacturers in the successful commercialization of autonomous solutions by aligning technology development, ecosystem configuration, and business model design."

Discussion

"The study contributes through the development of a maturity framework for autonomous solutions that captures the dimensions of technology development, ecosystem configuration, and business model design across three levels of maturity."

"By highlighting the interplay between technical-system evolution, ecosystem configuration, and business model design, the framework supports managers in taking a holistic perspective when making decisions concerning the development of autonomous solutions."

"Studying the correlation between the maturity level of the autonomous solution and the business model design in other contexts could provide additional insights that enable a more generalizable framework to be produced."

"Our study lays the foundation for further research that seeks to develop a process understanding of business model innovation, capturing the phases and activities that industrial equipment manufacturers undertake when transitioning to increasingly advanced autonomous solutions."

Artificial Intelligence for Decision Support Systems in the Field of Operations Research: Review and Future Scope of Research [80]

This is a machine-generated summary of:

Gupta, Shivam; Modgil, Sachin; Bhattacharyya, Samadrita; Bose, Indranil: Artificial intelligence for decision support systems in the field of operations research: review and future scope of research [80]

Published in: Annals of Operations Research (2021)
Link to original: https://doi.org/10.1007/s10479-020-03856-6

If you want to cite the papers, please refer to the original.

For technical reasons we could not place the page where the original quote is coming from.

Abstract-Summary

"Operations research (OR) has been at the core of decision making since World War II, and today, business interactions on different platforms have changed business dynamics, introducing a high degree of uncertainty."

"Our study reviews and investigates the existing research in the field of decision support systems (DSSs) and how artificial intelligence (AI) capabilities have been integrated into OR."

"The findings of our review show how AI has contributed to decision making in the operations research field."

"This review presents synergies, differences, and overlaps in AI, DSSs, and OR."

"Past studies were calibrated according to prognostic capability, exploitation of large data sets, number of factors considered, development of learning capability, and validation in the decision-making framework."

"This paper presents gaps and future research opportunities concerning prediction and learning, decision making and optimization in view of intelligent decision making in today's era of uncertainty."

Introduction

"AI's power to create near-to-real, true business scenarios and decisions makes it most appropriate in its application by organizations as a game disrupter [81]."

"Combined use with AI needs to be embedded in OR for real-time and optimal decision making."

"AI utilizes neural network models for the decision-making process [82]."

"DSS-based AI, GIS, and remote sensing have been applied for effective public decision making [83]."

"These studies utilized different capabilities of AI that were suitable in their scope, but questions still remain regarding the kind of AI capabilities that have been used to date to support decision making in different scenarios."

"In an uncertain era for many businesses such as today, it is more appropriate to use the prognostic decision-making capabilities of AI."

"RQ2: How are AI capabilities such as prognostic decision making used to address complex problems in uncertain environments?"

Background of the Study

"These models help firms make the best decisions and maximize their profits [84, 85]."

"AI offers an extremely low error rate in decision making compared to humans and other systems."

"AI tools that assist DSSs are referred to as knowledge-based systems, expert systems, joint cognitive systems, and intelligent decision systems [86–88]."

"When combined with DSSs, AI tools such as artificial neural networks, case-based reasoning, machine learning, cognitive computing, probabilistic reasoning, genetic algorithms, fuzzy theory, and multi-agents can help in fast decision making to evaluate and select the best alternatives [89, 90]."

"These fields can be supported with the help of AI techniques ranging from case-based reasoning, fuzzy logic, knowledge-based systems, genetic algorithms, and hybrid techniques [91, 92]."

"AI leads us towards decision making through DSSs."

Research Methods

"To identify the keywords in this review, we adopted a structured process of independent searches, first through Scopus on AI, DSSs, and OR, respectively."

"We also planned to limit the search to "article, article in press, review paper; business management, accounting, and decision sciences field"."

"In the sixth stage, the search was limited to journal papers, journal papers in press, and review papers in the business management, accounting, and decision sciences domains."

"The Annals of Operations Research (ANOR), the European Journal of Operational Research (EJOR), and Decision Support Systems-like journals are dedicated to OR and related practices in decision making."

"This section summarizes the literature review based on the following parameters: (1) year of publication, (2) publication source, (3) institution involved in the research, (4) methodological approach, and (5) underlying theory/theories."

"Fuzzy-based studies found applications in the finance stream along with healthcare and knowledge management, but with fewer journal papers."

Key Functions of AI-DSSs in OR

"Research is required with regard to the following points: (1) How can trust be developed among decision makers using AI-facilitated decisions? (2) How can decisions be made with less machine processing power in business? (3) What about the integration of convolutional neural networks with AI for improved DSSs?"

"Multi-agent-based knowledge integration mechanisms can be assimilated to agent-based models for better decision making [93]."

"This study is adequate and unique due to its greater focus on the application of AI in DSSs and OR in a number of business operations."

"Decision making using support systems can help designate resources obtained from the surrounding environment as well as the firm's internal resources [94, 95]."

"The information retrieval capability of DSSs utilizing AI and OR concepts can help obtain practically the exact information required for decision making [96]."

Conceptual Framework and Propositions

"With the help of the OR fundamentals ingrained in AI systems, firms can be in a better position to decide on an appropriate response when balancing the economic, environmental, and social aspects of sustainable performance."

"AI-coupled OR systems can help develop and train employees to reducing bias and improve transparency in decision making."

"Apart from making sure that productivity is on track, AI-driven DSSs can help facilitate strategic investments by bypassing multiple agents in payment systems and ultimately reducing the cost of operations [97]."

"The combination of AI, DSSs, and OR can help firms to better design, develop, and allocate their transactions (transaction cost economics), resources (resource based view), and capabilities (dynamic capability view) along the supply chain."

"The intelligent DSS works using the elements of OR and AI and helps businesses make speedy and optimal decisions that are driven by distributed and resilient systems to achieve a competitive advantage for the organization."

Discussion

"We developed a systems framework that can lead to an intelligent decision-support system via the integration of AI and OR."

"Managers may also help researchers with available data to identify and develop suitable frameworks for their decision-making processes."

"Business decision making has been emphasized through potential AI applications, but studies that discuss its societal needs in decision making scenarios are lacking."

"RQ2: How are AI capabilities such as prognostic decision making used to address complex problems in uncertain environments?"

"When information is lacking, AI utilizes network agents to model sequential decision making in the uncertain environment [98]."

"AI systems look for new approaches to train with the data and make decisions that are closest to reality."

"In order to make effective decisions, AI systems in uncertain environments seek more data and learn how to navigate confusing situations."

Conclusion and Future Research Directions

"We conducted a review of AI, DSSs, and OR and the features that allow them to help businesses."

"Following the three-step approach, we presented the key findings and developed propositions resulting in a conceptual framework that defines the impact and features of AI, DSSs, and OR."

"The reviewed studies have developed a number of models for decision making."

"Inter-operability of different kinds of data in real-time decision making is another aspect of future research where AI can be applied in combination with OR to achieve a sophisticated DSS."

"Not many studies have been able to integrate AI and OR [99]."

"Further research needs to be conducted to map consumers in a retail setting to offer complimentary products via AI and OR."

"OR could also be enhanced in terms of deep learning with the assistance of quantum computing and then combined with AI for effective DSSs."

Cross Reality to Enhance Worker Cognition in Industrial Assembly Operations [100]

This is a machine-generated summary of:

Simões, Bruno; De Amicis, Raffaele; Barandiaran, Iñigo; Posada, Jorge: Cross reality to enhance worker cognition in industrial assembly operations [100]

Published in: The International Journal of Advanced Manufacturing Technology (2019)
Link to original: https://doi.org/10.1007/s00170-019-03939-0

If you want to cite the papers, please refer to the original.
For technical reasons we could not place the page where the original quote is coming from.

Abstract-Summary
"We discuss how Cross Reality (XR) and the Industrial Internet of Things (IIoT) can support assembly tasks in hybrid human-machine manufacturing lines."

"We describe a Cross Reality system, designed to improve efficiency and ergonomics in industrial environments that require manual assembly operations."

"In the design of our XR system, we discuss how aspects of content creation can be automated for short-lived tasks and how seamless interoperability between devices facilitates skills transfer in human-machine hybrid environments."

Introduction

"Machines can impair workers' performances and well-being when automation is poorly designed, and the fact that human workers are more prone to errors than machines also poses a great challenge in the design of new assembly lines and automation processes."

"Maintaining the presence of human workers in the assembly chain can create a competitive advantage because humans can rely on their natural senses to form complex and intuitive, yet instant, solutions whereas robots require reprogramming to address new product families and manufacturing problems."

"Manufacturing industries have found that augmented reality (AR), virtual reality (VR), and mixed reality (MR) designs and technologies have the potential to empower workers, actively and passively, to perceive and learn information about new assembly processes in a timely manner, without altering their established work routines."

"Our objective is to empower human workers with means to augment their production environments with information that adapts to new product families and facilitates the acquisition of new assembly skills."

Literature Review

"The use of AR, VR, and MR to enhance natural environments and human cognition has been an active research topic for decades."

"Sand and others [101] developed a prototype to project instructions into the physical workspace, which enabled the worker not only to find the pieces but also to assemble products without prior knowledge."

"Another major problem facing IAR is the development overhead of AR applications, which requires the creation of content and the design of the worker experience."

"Competitive advantages of AR and VR in industrial assembly and maintenance have been demonstrated in several studies."

"Boud and others's [102] experiments demonstrated that task completion times were longer when using the 2D drawings to train how to assemble water pumps before assembling the real product, in comparison with AR and VR training."

"Other advantages presented by VR, AR, MR, and XR to assembly scenarios include reduction of data retrieval times [103], improvement of ergonomic behaviors in assembly [104], more comprehensive skill transfers, and reduction in training times [105–108]."

Methodology

"In the classic workflow, worker disabilities affect their ability to perform their job at three levels: interpretation of the wiring instructions to be executed, performance of conductivity checks on components assembled, and interaction with robots."

"Assembly tasks present a unique set of challenges for XR, ranging from interaction issues between workers and robots to human factors and production requirements."

"In the first phase, which we call the design phase, we define what the task is and how assembly instructions can augment the worker cognitively."

"The system should adapt the visual context and schedule a new set of actions whenever a worker undertakes an action (e.g., installing a component) that deviates from these instructions without affecting the assembly sequence."

"A worker-driven XR application has to adapt the assembly instructions to different task contexts and worker skills, requiring a significant level of system modularity."

Proposed Methodology and System

"Our system relies on a Rule Management System (RMS) to enforce condition-based decisions that use technology installed on the shop floor to provide workers with an augmented assembly experience that seamlessly spans over disparate interaction devices, namely mobile tablets and Microsoft's HoloLens."

"In order to allow the seamless interaction between machines in the physical environment, we needed to implement a communication bridge for each sensor, PLC, device, and robot on the shop floor."

"The implementation of digital twins for different machines enables designers to run simulations that demonstrate integrated systems prior to their deployment and to predict the time required to their installation."

"To providing information on the state of the device, the digital twin also works as an interaction API, e.g., request the robot to pick an object in a given position by inputting new values for each sensor composing the robot in the digital twin."

Evaluation

"Since immersive approaches for human-robot interaction are new enough not to have an established baseline, our evaluation focuses on the following goals: (1) to assess the usability of our system, (2) to collect qualitative feedback on the design of the system, and (3) to record how people interact with the remote system via the provided user interface (UI)."

"The questions we seek to answer with this study are: (RQ1) Can new and infrequent users use and effectively exploit the features of the authoring tool to create spatially augmented assembly stations? (RQ2) To what extent are users capable to understand and model the interaction for the worker skills?"

"The prototype was designed to draw worker attention to relevant information about the task (e.g., highlight components, depict symbols to identify error-prone steps, to associate aural notifications to different states of the task), and to provide step-by-step guidance with media (e.g., 3D animations, sounds, colors, videos, and interactive checklists)."

Observations

"The perception of a well-integrated system with their physical space and tasks goals rated as the highest user score (average score of 4.71)."

"Participants also reported that information has easily accessible and consistent with their needs (average score of 1.43)."

"The entire system was a novelty to the workers, nevertheless the overall reaction is that they felt very confident while interacting with the system (average score of 4.43)."

"In a post-interview, participants described the overall tool as intuitive and flexible."

"With regard to research question 2, participants manifested difficulty to autonomously redefine the proposed interaction model created by the system for the different skills and disabilities."

Discussion and Limitations

"Traditional methods to enhance the skills and abilities of the worker, such as on-the-job training (OJT), cannot fulfill the requirements that are expected in future trends for the assembly sector."

"The purpose of this work was to analyze the use of XR systems as a training and supervising medium for workers with disabilities in industrial settings."

"Traditional applications might not be well-designed enough to communicate across disparate networks, thus, reducing their interaction potentials to a limited number of sensors and devices."

"Work, we need to investigate how to maintain the conversation active between the system and workers with cognitive disabilities, e.g., prompting information about the current action and reminding the worker how to interact with the system once the action is completed."

"The system has shown promising results in teaching and supervising workers with disabilities that otherwise would require human supervision."

Conclusions and Future Work

"Our system is designed to augment, in a visual manner, human workers with the information necessary to undertake specific, frequently changing assembly tasks."

"Future research will aim to better understand from a cognitive perspective how humans interact with virtual objects in a XR environment so as to be able to design a system which effectively and appropriately augments human performance through the detection of cognitive needs rather than through users' deliberate actions."

"The central hypothesis is that relevant cognitive bottlenecks can be systematically identified and addressed by an intelligent cognitive architecture that receives input from XR hardware, biometric sensors, workplace data monitors, and intelligent machinery, and that this information can be leveraged to achieve effective human augmentation."

"The rationale behind conducting this future research is that a detailed knowledge of the possible modes of user interaction must be fully understood in order to allow for the design of the most effective XR infrastructures with our system."

Role of Artificial Intelligence in Rotor Fault Diagnosis: A Comprehensive Review [109]

This is a machine-generated summary of:

Nath, Aneesh G.; Udmale, Sandeep S.; Singh, Sanjay Kumar: Role of artificial intelligence in rotor fault diagnosis: a comprehensive review [109]
Published in: Artificial Intelligence Review (2020)
Link to original: https://doi.org/10.1007/s10462-020-09910-w

If you want to cite the papers, please refer to the original.
For technical reasons we could not place the page where the original quote is coming from.

Abstract-Summary
"Rotor faults have drawn more attention from the AI research community in terms of utilizing fault-specific characteristics in its feature engineering, compared to any other rotating machinery faults."

"Within this scenario, this paper is the first of its kind to attempt to review and define the role of AI in RFD and provides an all-encompassing review of rotor faults for the researchers and academics."

"This study is unique in three ways: (i) it emphasizes the use of fault-specific characteristic features with AI, (ii) it is grounded in fault-wise analysis rather than component-wise analysis with appropriate fault categorization, and (iii) it portrays the current research and analysis in accordance with different phases of an AI-based RFD framework."

Introduction

"In the early days, the rotor fault diagnosis (RFD) research was categorized into two main streams: (i) model-based diagnosis and (ii) signal-processing-based diagnosis."

"More research studies on the model-based identification of rotor faults can be found in the literature review from Bachschmid et al. [110]."

"The majority of the research related to AI-based fault diagnosis of rotating machinery only focuses on the bearing or gear faults [111–113], with little attention paid to RFD."

"The majority of the studies on rotating machinery fault diagnosis focus on component-wise analysis rather than fault-wise analysis, thus leaving loopholes in utilizing the fault characteristics and the associated features [114]."

"While structural rotor faults (SRF) are a common and straightforward fault in rotating machinery, compared to bearing or gear faults, they receive little attention from the research community."

AI-Based RFD Framework

"There are individual attempts [115] made to perform the comparative analysis of the above algorithms using the same dataset and signal pre-processing methods."

"CNNs [116] have attracted more attention than any other DL model in RFD since they have proven to be as effective with temporal data as with images, leaving the user with the responsibility of providing appropriate input representation."

"The sequential DL models, well known for their ability to deal with the temporal data, such as a recurrent neural network (RNN) and the variants, known as long short-term memory (LSTM) [117] and gated recurrent unit (GRU) [118], have also been explored in relation to RFD."

"The deep generative model known as the generative adversarial network (GAN) [119] has performed its role as a data upsampler in certain RFD works."

Theoretical Background of Rotor Faults

"In terms of the couplings, misalignment creates excessive heat and friction, and thereby damages the component, while in terms of the horizontal shafts, any misalignment causes excessive vibration on both the vertical and axial planes."

"In terms of vertically placed shafts, the unbalance is due to the mass effects of radial plane vibration, and the dominant frequency component will be 1x."

"A bent shaft is the most common rotating machinery fault and often develops as a result of thermal distortion, creep, or a large unbalance force [120], while rotor rub is another cause of thermal rotor bowing."

"Fault characteristics analysis: The cracked shaft vibration mostly affects the radial plane, and it produces increased 1× vibration along with the 2× and 3× harmonics."

"Excessive unbalance, misalignment, self-excited instability, and resonance can cause stator-rotor rub faults in the system, which results in highly non-linear vibrations [121]."

Machine Learning-Based Approaches to RFD

"Another work from the perspective of performance comparison was done by Tajik et al. [122], who applied six different methods for feature extraction and three different NNs (BPNN, RBFNN, PNN) for classification with PCA and LDA dimensionality reduction techniques, for different rotor unbalance conditions."

"In another similar work, Moosavian et al. [123] compared the performance of the SVM and K-NN on unbalance rotor fault with 29 FD features and found that though KNN is faster than SVM, the latter shows substantially better performance in terms of accuracy."

"Godoy et al. [124] employed fuzzy ARTMAP network, SVM, k-NN, and MLP to classify broken rotor bar defects in inverter-fed induction motors analyzing the TD stator current signal, and the rate of acquisition was based on machine frequency supply."

"Li et al. [125] has worked with WPD based feature extraction for classifiers BPNN, SVM, and k-NN on looseness, friction, and load unbalance faulty conditions."

Deep Learning-Based Approaches to RFD

"The proposed CNN contained 32 convolution filters and 200 hidden units in the fully connected layer with scaled DFT of raw data as input, and it achieved an accuracy of 93.61% over 87.25% of the conventional model."

"Of converting raw data (that are sampled into a phonetic form) to 2-D image input for the learning of CNN, based on the spectrogram, was proposed by Yu et al. [126]."

"Yongbo et al. [127] developed a method using CNN that extracted fault features from the IR thermal (IRT) images and fed it into the softmax regression classifier for fault pattern identification."

"The proposed multi-input convolutional neural network (MI-CNN) took the raw TD input signal and applied to 1-D CNNs, in which kernels of different sizes were adopted to extract the signal features of diverse dimensions."

"For integrating automatic feature learning, adaptive feature fusion, and accurate fault classification of data from multiple sources, a multi-stream CNN was proposed by Yuan et al. [128]."

Summary and Challenges

"Data acquisition phase: Methods in both ML and DL pursue the same process in the data acquisition phase of RFD, though they demand data at different quantity and precision."

"This limits the scope of RFD specific research improvements, which in turn causes a proportional falling off in the percentage of AI works that utilized rotor fault specific parameters (DFC, frequency harmonics, etc) in ML with 14% and DL with 8%."

"Certain signal processing techniques have been utilized in DL models for data preprocessing and proper input representation."

"It is evident that DL methods for RFD also demand certain signal processing with 26% and 18% in the FD, and TFD, respectively."

"Since the raw signals acquired from sensors are affected by noises, it is observed that the researchers depended on preprocessing steps to deal with noisy raw data, rather than going for NB like ML models which project staunch robustness to noise."

Future Research Directions

"This poses a challenging research problem in time-series domain and augmentation literature."

"GAN based data generation has been tried in certain bearing fault diagnosis works, leaving scope for research in RFD data generation."

"Challenging datasets: Another research lies in filling the gap between testbed data and real-world industrial data."

"Challenging Problems: Most of the RFD research deals with single faults, and only very few have tried to study compound faults, which is not on par with the real-world scenario."

"The research challenge in dealing with compound faults can be made more complicated by considering the compound faults as a multi labeling problem."

"The training with artificial faults and the transfer of the trained model for the classification of natural faults in a real-world scenario (by facilitating classifier fusion, transfer learning, and domain adaption) is a challenging problem in research."

Conclusion

"This paper focused on conducting a comprehensive review of the literature related to RFD involving ML and DL strategies."

"RFD was further discussed in relation to an AI-based framework to present the methodologies adopted in each phase, demonstrating the pros, cons, and challenges to provide an overview of the current RFD scenario."

"The review on the ML and DL classifier models was presented with the aim of emphasizing the rotor fault specific factors, such as the ability to incorporate fault-specific symptom parameters in feature engineering and dealing with imbalanced, scarce, or noisy data."

"This work is presented to provide a single point of reference for RFD researchers in relation to AI with practical insights."

Vision-based real-time monitoring of extrusion additive manufacturing processes for Automatic Manufacturing Error Detection [129]

This is a machine-generated summary of:

Charalampous, Paschalis; Kostavelis, Ioannis; Kopsacheilis, Charalampos; Tzovaras, Dimitrios: Vision-based real-time monitoring of extrusion additive manufacturing processes for automatic manufacturing error detection [129]

Published in: The International Journal of Advanced Manufacturing Technology (2021)

Link to original: https://doi.org/10.1007/s00170-021-07419-2

If you want to cite the papers, please refer to the original.

For technical reasons we could not place the page where the original quote is coming from.

Abstract-Summary

"The scientific fields of additive manufacturing and especially the extrusion-based technologies have gained immense attention in numerous commercial and research areas in the past decades."

"Monitoring the manufacturing procedure and detecting errors during the process remain a technological challenge in the field."

"Defect detection and dimensional accuracy inspection of the produced component is applied after the manufacturing has been completed and is accomplished via on-site manual monitoring."

"To overcome these issues, the present paper introduces a vision-based method to scan, filter, segment, and correlate in real-time the physical printed part with the digital 3D model as well as to evaluate the performance of the additive manufacturing process."

Introduction

"In-situ monitoring (ISM) of the AM procedure could diminish the feedstock material resources and manufacturing time due to the fact that the building process would be aborted in cases of a malfunction or an error is detected."

"These tools would be useful in AM procedures in order to increase the productivity of the process as well as to ensure a minimum quality on the manufactured part."

"Towards this direction, FDM processes can enhance their reliability and robustness by employing a generalized quality control technique that computes dimensional deviations between the manufactured model and the digital one."

"The implementation of the method was accomplished via the integration of a 3D structural light scanner on an FDM 3D printer and the development of a vision-based registration algorithm in order to detect dimensional deviations during the manufacturing process."

Related Work

"The technique was based on an optical measurement method that is able to capture stereoscopic images during the process and correlate them spatially extracting that way the external surface of a printed component."

"The state of the printing process was classified as "defective" or "good" applying a trained machine learning model and the images captured during the building procedure."

"The majority of the research work conducted so far is concentrated in examining certain types of failure mechanisms and is very sensitive to the hardware setup, whereas in the present study, several types of defects can be detected via the calculation of the dimensional deviations between the digital and the physically printed model using a 3D scanner."

"More particular, a fully automatic monitoring and error detection system was developed utilizing a high-resolution structural light 3D scanner, where the dimensional deviations between the digital 3D model and the printed one are computed during the AM process."

Methodology

"In order to be able to calculate the dimensional deviations between the physical printed part and the digital 3D model, it is necessary to convert the G-code, which is sent to the 3D printer, to an equivalent 3D point cloud that contains information about the internal and the support structures of the printed object."

"The utilized method considers the support as well as the internal structure of the product in order to construct the real-theoretical point cloud of the digital 3D model."

"Due to the fact that the build platform of the employed 3D printer is stationary during the AM process, the model's reconstruction can be accomplished via the concatenation of each individual filtered point cloud of every scanned layer."

"The segmentation of each printed layer and the theoretical point cloud, as well as their transformation was achieved enabling that way the possibility to compare and calculate their differences during the manufacturing process."

Results and Validation of the System

"In the first row of the figure, the dimensional deviations between the reconstructed and the theoretical point clouds are shown in several steps of the AM process."

"The results show that as the AM process progresses, the dimensional deviations also grow leading finally in a MAE and RMSE equal to 0.223 and 0.331 accordingly."

"Hereupon, the results validated the efficiency of the suggested system and confirmed that is capable of real-time monitoring the FFF AM process."

"This work presents an application of the introduced real-time monitoring and error detection system demonstrating its capabilities in evaluating the quality of a 3D printed part with complex geometrical shapes."

"The first and second rows present the theoretical and the reconstructed point cloud correspondingly, and the third raw documents the dimensional deviations between the above-mentioned clouds."

Conclusions

"The suggested online system is automatically operated exploiting 3D data measurements of the printed part, point cloud processing, file transportation, and decision-making strategies coupled with correlation between the digital and the printed product."

"In the last step, the proposed system computes the dimensional deviations between the theoretical 3D point cloud and the corresponding printed model and makes a decision to continue or abort the AM process based on the calculated results."

"It is deduced through the conducted experiments that the introduced methodology and the applied vision-based algorithms can efficiently capture and analyze the geometrical characteristics of the manufactured component during the AM process."

"Based on the technology that was developed in the present study, future work can be conducted to automatically repair the printed part via modifications on the values of the employed printing parameters during the AM process."

Analogue Models and Universal Machines. Paradigms of Epistemic Transparency in Artificial Intelligence [130]

This is a machine-generated summary of:

Greif, Hajo: Analogue Models and Universal Machines. Paradigms of Epistemic Transparency in Artificial Intelligence [130]

Published in: Minds and Machines (2022)
Link to original: https://doi.org/10.1007/s11023-022-09596-9

If you want to cite the papers, please refer to the original.

For technical reasons we could not place the page where the original quote is coming from.

Abstract-Summary

"The problem of epistemic opacity in Artificial Intelligence (AI) is often characterised as a problem of intransparent algorithms that give rise to intransparent models."

"The degrees of transparency of an AI model should not be taken as an absolute measure of the properties of its algorithms but of the model's degree of intelligibility to human users."

"In order to elucidate this claim, I first contrast computer models and their claims to algorithm-based universality with cybernetics-style analogue models and their claims to structural isomorphism between elements of model and target system (in: Black, Models and metaphors, 1962)."

"I conclude that their respective degrees of epistemic transparency primarily depend on the underlying purposes of modelling, not on their computational properties."

Introduction

"In areas where users or citizens in general might be negatively affected by intransparent operations of AI, epistemic opacity is framed as a normative problem that should be resolved, regulated or altogether avoided."

"In the domain of scientific modelling though, epistemic opacity assumes a remarkably ambiguous status, variously figuring as an unwelcome constraint that can and should be overcome or as an 'essential' epistemic condition that affects computer models more generally."

"This paper is an attempt to resolve some of the ambiguity in the notion of epistemic opacity in the context of AI-based scientific modelling."

"This analysis takes issue with the dichotomy between the view that epistemic opacity is chiefly a matter of complex or otherwise intractable algorithms that can be resolved on the algorithmic level and the claim that opacity is an 'essential' characteristic of AI models."

Parsing Epistemic Opacity

"The key to Humphreys' analysis of epistemic opacity as being 'essential' in a relevant class of cases is his twofold assumption that, in the use of computer models in science, the goal of epistemic transparency is unattainable for human observers in principle, and that it is unattainable in principle by virtue of some of the specific characteristics of the computational method."

"Of non-essential opacity, an opaque model can be made more transparent if its complexity can be reduced, if it can be decomposed into its operationally pertinent elements, if the elements' behaviours can be reliably predicted, or if one can devise other means of interpreting or meaningfully modelling the properties relevant to the model's interpretation in a given epistemic context."

"The twofold premiss of this paradigm is that reducing or modelling computational complexity or related factors in view of a human-understandable solution is possible in a given context of application, and that the algorithms or other computational elements involved are in fact the epistemically relevant elements in that context."

Analogue Models and the Cybernetic Paradigm

"It is a symbolic representation of some real or imaginary original, subject to rules of interpretation for making accurate inferences from the relevant features of the model (Black [131], p. 222, emphasis in original). More precisely, analogue models establish isomorphism relations, understood in the mathematical sense as 'point-to-point correspondence' between features of the model and features of the original or, in more contemporary parlance, target system."

"Not every relation between model and target system that would formally qualify as isomorphic or structurally corresponding will be included in the model but only those which allow for 'independent confirmation', given that analogue models can only 'furnish plausible hypotheses, not proofs' [131], p. 223."

"The epistemically relevant elements of Phillips' MONIAC, Ashby's homeostat or Walter's tortoises are the valves, tubes, electric circuits and other material components whose operations and relations within the model are designed to partly represent and contribute to an explanation of the structure and behaviour of their respective target systems."

Computer Models and Universal Machines

"If an observer wants to learn more about what an algorithmic structure might represent beyond specifying a computation, he or she must move to higher-order properties of the model."

"These machines are humans who calculate.') At the other end of the spectrum, there is the claim that all relevant processes that can be modelled by computers are computational processes themselves, so that the model's algorithms reflect the fundamentally computational structure of its target system."

"A claim for a shared set of functional principles between model and target systems is thereby established at the algorithmic level, but at the cost of inflating the extension, and possibly also the meaning, of the concept of computation."

Case Studies of Model Intelligibility

"Whereas, for example, the recourse to back-propagation mechanisms in classical connectionism will constitute a negative analogy to the functions and structure of natural nervous systems and make its models less realistic in light of their cognitive modelling aims, DNNs are less affected by this limitation to the extent that they typically do not seek to provide cognitive models, let alone propose shared mechanisms, in the first place."

"Operating, like the majority of modern cognitive science, in a tradition that takes the problem of perceptual ambiguity as its starting point, it posits that an organism has to infer the structure and properties of the world from ambiguous and underdetermined sensory input. (Berkeley [132] was the first to provide an explicit account of how three-dimensional images of an object or scene arise from a two-dimensional retinal image.) However, in contrast both to the notion of symbolic information processing in classical AI and to classical and DNN connectionism, PP claims to realistically model cortical information processing as a bidirectional hierarchical process of prediction error minimisation between sensory input and higher-order dynamic world models."

Conclusion

"The argument on the previous pages can be su4mmarised as follows: With respect to scientific modelling, epistemic transparency is a function of the cognitive grasp of the epistemically relevant elements of a model that its users and designers develop."

"Computer models face a particular difficulty with respect both to defining its epistemically relevant elements and to establishing a cognitive grasp of these elements."

"The model's algorithms neither necessarily nor naturally constitute its epistemically relevant elements."

"I have begun to develop an argument that computer models are not fundamentally different from DNNs in terms of epistemic opacity."

"DNNs, like ML approaches more generally, are neither designed for this endeavour, nor can they be recruited for it in a similar way to analogue or more traditional computer models."

Thirty Years of Artificial Intelligence and Law: The Third Decade [133]

This is a machine-generated summary of:

Villata, Serena; Araszkiewicz, Michal; Ashley, Kevin; Bench-Capon, Trevor; Branting, L. Karl; Conrad, Jack G.; Wyner, Adam: Thirty years of artificial intelligence and law: the third decade [133]

Published in: Artificial Intelligence and Law (2022)
Link to original: https://doi.org/10.1007/s10506-022-09327-6

If you want to cite the papers, please refer to the original.

For technical reasons we could not place the page where the original quote is coming from.

Abstract-Summary

"The first issue of Artificial Intelligence and Law journal was published in 1992."

"They indicate a major shift within Artificial Intelligence, both generally and in AI and Law: away from symbolic techniques to those based on Machine Learning approaches, especially those based on Natural Language texts rather than feature sets."

Introduction

"The most significant feature of the journal's third decade was the dramatic increase in the application of machine learning techniques to a variety of AI and Law tasks."

"None the less the rise of machine learning in the second half of the decade under consideration was striking, and the bulk of the papers discussed in this article relate to that development."

"The second, Boella et al. [134], commented on by Adam Wyner, describes a comprehensive suite of tools to provide for the management of legal documents on the Web."

"The remaining six papers discuss various aspects of the application of Machine Learning techniques to AI and Law."

"The paper suggest that by using machine learning to ascribe factors, use can be made of the techniques for explanation developed for traditional AI and Law systems."

"The other papers discuss a variety of other legal tasks."

[Section 2]

"The paper presents a model of statutory legal knowledge (provisions and related axioms) encompassing Hohfeldian relations, using the RDF(S)/OWL standard developed by the World Wide Web Consortium."

"The model enables the query and retrieval of particular types of provisions as well as reasoning over the stored knowledge using a Description Logic (OWL-DL)."

"The model addresses a practical problem concerning the retrieval of all related provisions from the statutory text."

"The Provision Model distinguishes between logical relations and technical relations between (the elements of) legal provisions."

"In order to represent these relations adequately, the provision attributes are specified to capture the provision type (for instance, the model will use the expression hasDutyBearer rather than hasBearer)."

"The proposed model enables advanced retrieval and reasoning over normative regulations (also in multilingual settings) by exploiting relations between legal provisions [135–137]."

[Section 3]

"Heterogeneous, distributed legal sources are in tension with user demands to access the law over the internet using Open Government Data and Linked Open Data."

"Approach: domain-specific ontologies of terms and the ontological concepts can be linked with text in the source."

"The core system of EUNOMOS has three layers: a database of Italian national laws that are processed for similarity (Cosine Similarity with TF-IDF) and for domain topic (SVM) then converted into XML using the ITTIG XML parser to convert the NormaInRete XML format; an ontology of legal terms from the Legal Taxonomy Syllabus, which are augmented with lexical semantic relations and contextual or interpretive information, then linked to the law (using the TULE parser [138]); and a user interface for querying the database with respect to the ontology, keywords, and other metadata."

[Section 4]

"Rather than proposing a normative model, explicating an argumentation framework, or depending on a manually curated representation, Abood and Feltenbergerer present a tractable solution to a significant real-world legal problem that integrates human expertise with recent advances in HLT and machine learning."

"Abood and Feltenbergerer uses a semi-automated approach that starts with a human-curated seed set that is then expanded through family citation and class code links into an over-inclusive set."

"Notwithstanding the applicability of this approach to legal data other than patents, citations to Abood and Feltenbergerer have been almost exclusively from patent analysis researchers, e.g., Choi et al. [139]."

"The lack of explanatory capability that makes machine-learning arguably inappropriate for discretionary legal judgements has far less relevance for data analysis tasks designed to facilitate, rather than substitute for, expert human judgments, provided that the predictive models' accuracy can be evaluated."

[Section 5]

"In Recurrent Neural Network-based Models for Recognizing Requisite and Effectuation Parts in Legal Texts, Nguyen and others train an RNN to identify and label two central components within Japanese legal documents: a "requisite part" and an "effectuation part" [140]."

"Nguyen and others propose several variations of neural network techniques for recognizing requisite and effectuation parts in legal text."

"They propose the sequence of Bi-LSTM-CRF models and two types of multi-layer models to recognize overlapping requisite-effectuation (RE) parts, which include the Multi-layer-Bi-LSTM-CRF and the Multi-layer-BiLSTM-MLP-CRF (the latter, multi-layer perceptron variant eliminates redundant components and can reduce training time and redundant parameters)."

"Given the two types of multi-layer models, the Multi-Bi-LSTM-MLP-CRF solves limitations of the Multi-Bi-LSTM-CRF because it eliminates redundant components, thus offering a smaller model and faster training and testing times with no sacrifice in performance."

[Section 6]

"We will discuss Medvedeva et al. [141], which was the first paper on ML prediction of legal cases to appear in the journal."

"Using the communicated cases rather than the facts from the final decisions led to a substantial (around 10%) degradation in accuracy in a series of experiments reported in Medvedeva et al. [142], leading to the conclusion that "the experiments conducted in this paper show that performance seems to be substantially lower when forecasting future judgements compared to classifying decisions which were already made by the court"."

"The second experiment in Medvedeva et al. [141] looked at a second problem for predicting cases."

"The work reported in Medvedeva et al. [141], and subsequent papers by this group represent a careful exploration of the possibility of classifying and predicting legal decisions using machine learning techniques unsupported by domain analysis."

[Section 7]

"Given a small sample of decisions annotated with legally relevant factual features, their program could predict outcomes of textually described cases, and identify the features that could help to explain the predictions, for example, by indicating the elements of the legal rule that have been satisfied or are still missing."

"To the extent that an approach like SCALE can identify findings, issues, and factors in other types of legal cases, it would help to connect modern text analytic techniques with knowledge-based computational models of case-based legal reasoning."

"If, in addition to predicting outcomes of textually described cases, text analytics could also identify the applicable factors, then case-based models that account for legal rules, underlying values, and case factors could assist machine learning in explaining and testing those predictions."

[Section 8]

"The paper addresses the task of creating a deep pre-trained language model for legal text, following the example of the popular BERT learning framework [143]."

"The authors define a model called LamBERTa (Law article mining based on BERT architecture) which fine-tuned an Italian pre-trained BERT on the Italian civil code text."

"To evaluate the effectiveness of the LamBERTa models for the law article retrieval task on the ICC, the authors conducted a comparative analysis with state-of-the-art text classifiers based on deep learning architectures: a bidirectional Long Short Term Memory (LSTM) model as sequence encoder (BiLSTM, Liu et al. [144]), a convolutional neural model with multiple filter widths for text encoding and classification (TextCNN, Kim [145]), a bidirectional LSTM with a pooling layer on the last sequence output ((TextRCNN), Lai et al. [146]), a Seq2Seq [147, 148] model with attention mechanism (Seq2Seq-A), and a transformer model for text classification adapted from Vaswani et al. [149]."

"To satisfy this need, the models were firstly provided with Italian Wikipedia pre-trained Glove embeddings, and secondly, each model was fine-tuned over the individual ICC texts following the same data labeling schemes as the LamBERTa models."

[Section 9]

"This paper presents an important contribution to the field of AI and Law by proposing a novel Natural Language Processing (NLP) approach to the identification of unfair clauses in a class of consumer contracts."

"The contribution of this paper is not limited to the novel approach to the identification and classification of unfair clauses in ToS documents."

"Two main contributions are presented in this paper: first, the creation of an annotated dataset for the identification and classification of unfair clauses in ToS documents, and second, the architecture designed to automatically address this task."

"Concerning the architecture to automatically identify and classify the ToS documents along with the set of five unfair clauses identified in the corpus, and keeping an eye on the transparency of the proposed approach, the authors decided to rely on a promising approach associating explanations to the outcomes of neural classifiers."

Service Selection Mechanisms in the Internet of Things (IoT): A Systematic and Comprehensive Study [150]

This is a machine-generated summary of:

Li, Yunyan; Huang, Yuansheng; Zhang, Meimei; Rajabion, Lila: Service selection mechanisms in the Internet of Things (IoT): a systematic and comprehensive study [150]

Published in: Cluster Computing (2019)
Link to original: https://doi.org/10.1007/s10586-019-02984-4

If you want to cite the papers, please refer to the original.

For technical reasons we could not place the page where the original quote is coming from.

Abstract-Summary

"Internet of Things (IoT) describes embedded devices (things) with Internet-based connectivity, enabling them to receive and send data through the communication network."

"The main aim of Service-Oriented Architecture (SOA) is to select the best services among a pool of services."

"These services can be selected statically or dynamically regarding the service functionalities and performance limitations."

"Since the performance of complex services is very important in many distributed domains."

"This study aims to systematically review the service selection mechanisms in the IoT. The service selection mechanisms are classified into centralized, decentralized, and hybrid classes."

Introduction

"In IoT, the selection of services and features among many similar services is a complex process [151–153]."

"There are three main challenges in the IoT service selection issue."

"In the context of the IoT, scalability is an important issue and must be considered in selecting the services [154]."

"Since the number of things and services on the IoT is increasing, selecting an appropriate service is very important."

"While a Systematic Literature Review (SLR) is a detailed strategy to recognize and study all related works about a particular area, this paper carries out an SLR-based approach for reviewing the selection mechanism in the IoT. The SLR builds a correct evaluation on service selection mechanism using trusted and exact papers."

"The main aims of this paper are: Presenting an SLR and studying the current approaches for service selection in the IoT. Describing the main challenges about the IoT service selection."

Related Work

"Bouzary and Chen [155] have surveyed service selection and composition methods in cloud manufacturing."

"Yang and others [156] have surveyed and classified service selection mechanisms based on user preferences and deployment in a multi-cloud environment."

"The IoT and service selection are not discussed in this paper."

"They also have discussed an architecture of cloud services composition based on SOA architecture."

"Asghari and others [157] have focused on service composition methods that were applied in the IoT environment."

"Service composition methods in the selected studies are classified into two main categories with a focus on functional and non-functional properties aspect."

"The paper is not related to the service selection in the IoT. Finally, new papers have not been discussed."

"The main features and advantages of each cloud service composition technique are also illustrated."

"The paper is not related to the IoT and service selection."

Background

"To all of these advantages, IoT has some limitations and challenges, such as naming and identity management, standardization and interoperability, privacy, protection and security, confidentiality, green IoT, service management, and network-based issues [158, 159]."

"The service selection architecture has three parts: job submission, scheduler and execution monitor [160–162]."

"Of service selection techniques in the IoT, some QoS criteria will be used, such as reliability, processing time, execution time, availability, stability, response time, flexibility, user preference, accuracy, scalability, performance, load balancing, and security issue."

"Execution cost: It is a pay for using a service to get a result by requester [163]."

"User preference: It is some of QoS metrics that the users want to receive the requested service based on those metrics."

"Security: Security issues refers to the policies, applications, and controls utilized to protect data, services, and the infrastructure of the IoT environment [163]."

Research Procedure

"It is utilized to achieve a comprehensive study of the service selection mechanisms in the IoT. The present research goals are recognizing and analyzing all related papers that have examined in service selection in the IoT. Another aim of this paper is understanding the main issues and challenges in the field of IoT service selection, which includes reliability, processing time, execution time, availability, stability, response time, flexibility, user preference, accuracy, scalability, performance, load balancing, and security issue."

"This paper addresses the following Research Questions (RQs): RQ1: What is the position of the service selection component in the IoT?"

"The process of papers selection has three main steps [164–168]."

"Google Scholar is employed as the main search engine to find related papers based on "allintitile" phrase and some keywords, including (service selection, IoT, internet of things, service select)."

Service Selection Mechanisms in IoT

"In the proposed method, to provide a dynamic service selection process, authors have presented a game-based model where each component can make a service selection decision independently."

"This method does not consider some QoS metrics, such as security, availability, user preferences and etc Finally, Kanagaraju and Nallusamy [169] have studied an IoT architecture for industrial applications and have presented a Register Service Selection (RSS) method based on security architecture to collect the essential data and transmit it to a cloud-based backend."

"Rapti and others [170] have proposed a decentralized IoT service selection method based on Artificial Potential Filed (APF), which are formed upon each user request to provide the services."

"With regard to IoT applications that can be centralized and decentralized, the hybrid methods should be considered for increasing the performance of the service selection methods by employing the benefits of both of them."

Discussion

"The selected papers are categorized into three categories that 5 of the 17 papers are centralized, 8 of them are decentralized, and 4 of the 17 are hybrid techniques."

"The decentralized methods do not have the centric manager that manages requests and responses, therefore, it is difficult to know which node failed, so each node must be pinged for availability checking."

"According to the surveyed papers, centralized mechanisms do not consider reliability, processing time, load balancing, response time, and user preference."

"The decentralized and hybrid mechanisms, unlike the centralized mechanisms, have considered reliability, processing time, load balancing, response time, and user preference."

"Security issues are not important in the decentralized and hybrid mechanisms, therefore, these mechanisms do not consider security issues comprehensively."

"In the hybrid mechanisms, the accuracy parameter does not have considered."

"Service availability is the most important issue in the service selection mechanisms, but the authors do not attend this parameter well."

Open Issue and Future Works

"There is not any mechanism, which solves all challenges related to service selection in IoT. For example, some techniques consider reliability, scalability, load balancing, routing performance, response time; but some considered processing time, execution cost, execution time, user preferences."

"The comprehensive analysis of the most important challenges have been discussed as follow: Response time: communication between IoT smart devices can effect on the response time of service discovery and selection procedures to lessen the delivery time for final selected services."

"The privacy issue is another main concern in IoT environment, which emphasizes on the protection of both users' and service providers' privacy and security precedencies."

"The main challenge of this matter is that a protection model for users or service providers' privacy preferences should be offered for contribution for protecting the critical information in the service selection process."

Conclusion and Limitation

"According to the SLR in service selection mechanisms from 2014 to 2018, the number of published papers is determined very high in 2017 and 2018."

"The results have shown that most of the papers try to improve response time, reliability, execution time, scalability, reliability, but user preferences, security and privacy issue are not considered in many papers."

"The results have shown that most of the papers tried to improve response time, reliability, execution time, scalability, and reliability, but, user preferences, security and privacy issue are not considered in many papers."

"The decentralized and hybrid mechanisms have been able to reduce response time, and pay more attention to user preference issues."

"The centralized mechanisms are appropriate for designing secure service selection mechanism."

"The centralized mechanisms have lower reliability, lower user preference and high processing time compared to the other techniques."

Metaheuristics: A Comprehensive Overview and Classification Along with Bibliometric Analysis [171]

This is a machine-generated summary of:

Ezugwu, Absalom E.; Shukla, Amit K.; Nath, Rahul; Akinyelu, Andronicus A.; Agushaka, Jeffery O.; Chiroma, Haruna; Muhuri, Pranab K.: Metaheuristics: a comprehensive overview and classification along with bibliometric analysis [171]

Published in: Artificial Intelligence Review (2021)

Link to original: https://doi.org/10.1007/s10462-020-09952-0

If you want to cite the papers, please refer to the original.

For technical reasons we could not place the page where the original quote is coming from.

Abstract-Summary
"Research in metaheuristics for global optimization problems are currently experiencing an overload of wide range of available metaheuristic-based solution approaches."

"Since the commencement of the first set of classical metaheuristic algorithms namely genetic, particle swarm optimization, ant colony optimization, simulated annealing and tabu search in the early 70s to late 90s, several new advancements have been recorded with an exponential growth in the novel proposals of new generation metaheuristic algorithms."

"We present a relatively new taxonomic classification list of both classical and new generation sets of metaheuristic algorithms available in the literature, with the aim of providing an easily accessible collection of popular optimization tools for the global optimization research community who are at the forefront in utilizing these tools for solving complex and difficult real-world problems."

Introduction

"There is the need for a proper documentation of all the metaheuristic algorithms developed so far with the motivation to analyze, categorize and synthesize them in a meaningful manner based on their design inspirational source, useful impact and application areas."

"Brownlee [172] presented a well-documented reference text that describes to an extent a large number of algorithmic techniques that covers bio-inspired computation, computational intelligence and metaheuristics in a complete, consistent, and centralized manner for the research community."

"In this current study, we present a more comprehensive compilation of approximately 300 different metaheuristic algorithms with their classifications based on design inspiration, variants, useful impacts and application areas."

"A systematic categorization of both classical and new generation metaheuristic algorithms with emphasis on design inspirational source, variants, classification, impacts, and application areas."

Background Work on Nature Inspired Metaheuristic Algorithms

"The purpose of this section is to clearly discuss the difference between the already published review on nature inspired metaheuristic algorithms and the present work."

"Xing and Gao [173] presents a rough guide to nature inspired metaheuristic algorithms."

"Over 200 nature inspired metaheuristic algorithms are scattered in the literature."

"The previous works mainly focuses on the core operations, applications and inspirational sources of the nature inspired metaheuristic algorithms."

"None of the previous works attempted bibliometric analysis of the comprehensive list of the nature inspired metaheuristic algorithms to show the impact of the algorithms in different domain despite the significance of bibliometric analysis in the literature."

Bibliometric Analysis

"Some other crucial research areas with bibliometric studies are: real-time operating systems [174], green supply chain [175], energy efficiency [176], etc. Lately, there have been several journal specific studies also i.e. the bibliometric analysis on a particular journal."

"We provide the publication and citation structure in metaheuristic research domain over the last 30 years."

"With number of publications, there were only 12 citations of the work related to this area."

"The link between the entities corresponds to the strength between them either in terms of number of publications, common references or co-citations."

"The red color cluster of Kaveh, Yang, Deb, Coelho, Das, Soto, and Gandomi are likely working on same area (Definalty "metaheuristic") and citing the same source in their reference lists."

"There are two prominent nodes in cluster 2 with green color i.e. Computers and operations research and European Journal of operational research."

Metaheuristics Inspirational Source

"Many of the so-called metaheuristic algorithms have been tagged as nature-inspired, and this is because the algorithmic design concepts have been motivated by simply borrowing some sort of inspirational source from nature."

"The bio-inspired metaheuristic algorithms are more popular and common in the optimization literature, more so, these algorithms are commonly based on a metaphor of a typical biological process or interactions."

"In illustrating one of the bio-inspired metaheuristics techniques, we consider the symbiotic organism search, a population-based metaheuristic algorithm that was proposed in 2014 by Cheng and Prayogo [177]."

"Some common example of these metaheuristics include, simulated annealing, black hole algorithm, tabu search, intelligent water fall algorithm, wingsuit flying search, cultural algorithm, colonial competitive algorithm, harmony search, teaching–learning-based optimization, and so on."

Generalized Classification of Metaheuristic Algorithms

"Most of the well-known metaheuristics algorithms are nature-inspired, but few took the motivation from other processes such as physics/chemistry-based, social science, sports-based etc Nature-inspired metaheuristic algorithms are algorithms inspired by intelligent bahaviour of natural systems."

"After the introduction and success of GA, SA and a few other research works for solving complex problems, researchers also started studying various natural or manmade processes to make an efficient solver. (i.e. to take the inspiration for the efficient algorithm)."

"The evolutionary computation (EC) involved the type of nature inspired metaheuristic algorithms that falls within the reams of natural evolution processes and mechanism."

"The evolutionary algorithms family produce the pioneered nature inspired metaheuristic algorithm called genetic algorithm where the initial population represent the candidate solutions to the problem and the solution space is the environment."

"The physical algorithms (PA) are the types of nature inspired metaheuristic algorithms that are derived from the computational process and mechanism of physical systems."

Taxonomic Categorization of Metaheuristic

"The categorization provided in this space is driven by the algorithm design inspirational source, characteristics of search methods denoted as class and number of swarm or agents employed during the search process."

"Other factors considered for this categorization include existing variants of the algorithms and their application and citation impacts."

"The popularity achieved by these five standard algorithms can be attributed to the fact that each of these algorithm in one way or the other have concrete theoretical and mathematical analyses framework which supports their superior performances."

"Further, the theoretical and mathematical frameworks provide good reasons as to why these algorithms work well in real-world."

Metaheuristics Applications Areas

"Studies in the design and application of metaheuristics techniques has grown massively in the past three decades as a practical solution approaches to solving wide range of real-world optimization problems."

"More so, beside the fact that metaheuristics have scaled perfectly well in finding good quality solutions to many complex and NP-hard problems (Nondeterministic polynomial time-hard problems), they are also able to perform extremely very well in situations where the classical or traditional exact optimization methods would fail to deliver satisfactory results."

"One of the main strength of metaheuristic techniques that has drawn wide interest to the scientific community and industry practitioners alike is the fact that metaheuristic problem solving techniques are able to generate good quality solution in relatively much less time than their exact optimization techniques counterparts."

Glossary of Metaheuristics

"The following list presents a glossary of metaheuristic algorithms comprising of both classical and new generation algorithms."

"Earlier, these two class of metaheuristic are differentiated based on the decades in which the algorithm was actually developed."

"Each algorithm in this list has been discussed previous and therefore, only their names a presented here for easy summary."

Conclusion

"Some other related studies have equally attempted to create a complete taxonomy of all the published metaheuristics for the easy of identifying strengths, weaknesses and possible application areas of the global optimization metaheuristic methods."

"Further, we successfully presented a more comprehensive and relatively new taxonomic classification list of both classical and new generation sets of metaheuristic algorithms available in the literature, with the aim of providing an easily accessible collection of popular optimization tools for the global optimization research community who are at the forefront in utilizing these tools for solving complex and difficult real-world problems."

"We hope that the overall overview and the detailed classifications of all the available up-to-date state-of-the-art and new generation metaheuristic techniques presented in this study will inspire further novel research in the field of global optimization methods and subsequently provide better insight into future designs of more practicable metaheuristic algorithms that would be capable of handling complex and large-scale real-world problems."

Biological Computation and Computational Biology: Survey, Challenges, and Discussion [178]

This is a machine-generated summary of:

Chelly Dagdia, Zaineb; Avdeyev, Pavel; Bayzid, Md. Shamsuzzoha: Biological computation and computational biology: survey, challenges, and discussion [178]
Published in: Artificial Intelligence Review (2021)
Link to original: https://doi.org/10.1007/s10462-020-09951-1

If you want to cite the papers, please refer to the original.
For technical reasons we could not place the page where the original quote is coming from.

Abstract-Summary
"Biological computation involves the design and development of computational techniques inspired by natural biota."

"Computational biology involves the development and application of computational techniques to study biological systems."

"We aim at highlighting how nature has inspired the development of various algorithms and techniques in computer science, and how computational techniques and mathematical modeling have helped to better understand various fields in biology."

Introduction

"Computer scientists have relied on biological systems for inspiration to develop new, nature based techniques for solving difficult computational problems."

"This discussion motivates the global contributions of this work which are (i) an up-to-date review on nature inspired algorithms, referred to as biological computation, while describing how nature has inspired the development of different algorithms and computational techniques, (ii) an up-to-date review on how computational techniques and mathematical modeling have helped to study biological systems, and (iii) highlighting the impact of each of these two disciplines, i.e., biology and computer science, on each other and how they can guide and benefit each other to not only improve the understanding of biological processes but also to make notable advances in the design of algorithms."

Biological Computation

"For a specific bio-inspired algorithm, the different proposed algorithmic versions are similar in their basic approach and in making use of the bio-inspired concepts, but they mainly differ either in the way they represent the information or in the way they are structured, e.g., optimized version, additional components, etc. In what follows, we will give an overview of the basic bio-inspired techniques within the defined four main branches."

"Most research has studied the food source searching behavior and based on this several bee colony optimisation algorithms have been proposed to mainly solve optimization problems, whether combinatorial or numerical."

"There are several other applications of the bees' colony optimization algorithms that are based on this specific bees' behavior; applications to solve stochastic vehicle routing problems for instance or the routing in wired computer networks."

Computational Biology

"It includes, but not limited to, various analytical methods, mathematical modeling and simulation for a number of broad areas, including analysis of protein and nucleic acid sequence, structure and function, evolutionary genomics and proteomics, systems biology, population genomics, regulatory and metabolic networks, biomedical image analysis and modeling, gene-disease associations, and development and spread of disease."

"To such representations, by contrary to the requirement to reconstruct a single string, the formalization of genome assembly problem for graph-based models was shifted to strings reconstruction that are substrings of a DNA sequence, called contigs (also known as unitigs or omnitigs) Medvedev [179]."

"We will discuss the MSA phenomenon, various key components of phylogenomic analyses, genome rearrangements and how research in computational biology helped developing efficient tools for estimating multiple sequence alignments, phylogenetic trees, genomic rearrangements, etc Gene sequences evolve under processes that include events such as substitutions, insertions and deletions (jointly called "indels") that can change the length of the sequences, and must be accounted for comparative genomics."

Algorithms in Biology: Challenges and Discussion

"We mention Kappa Boutillier et al. [180] an integrated suite of analysis and visualization techniques for building and interactively exploring rule-based models, BioNetGen a software for generating mathematical/computational models that account comprehensively and precisely for the full spectrum of molecular species implied by user-specified activities, potential modifications and interactions of the domains of signaling molecules, BioSPICE Dashboard intended to assist biological researchers in the modeling and simulation of spatio-temporal processes in living cells, INSILICO discovery an advanced computational tool for network oriented "in silico" analysis and design of cellular properties, and many more."

"Although the biological computation community was making many efforts in developing a wide range of simulation systems for biology, the use of these software tools remained somehow restricted, and called for further improvements."

"If a computational biologist formulates the correct computational and mathematical problem, it would be much easier for a researcher in biological computation to understand which method should be used for solving this problem."

Conclusions

"We aimed at presenting how various computational techniques have been inspired from the nature, and, on the other hand, how computational techniques and mathematical modeling have been used to solve various biological questions."

"We presented a reasonably comprehensive review on biological computation, showing the tremendous impact and contribution this nature has on developing computational techniques."

"Research in biological systems and computational methods are converging and significantly contributing towards the advancement of science and technology."

"Neural networks and genetic algorithms are undoubtedly two of the most important computational techniques based on biological processes."

"This sort of collaborative efforts will introduce a nice balance between the "wet" lab and "dry" lab research which will ultimately result in an improved understanding of biological processes as well as fast and accurate computational techniques."

References

1. Mirra G, Pugnale A (2022) Expertise, playfulness and analogical reasoning: three strategies to train Artificial Intelligence for design applications. Arch Struct Construct 2(11):111–127. https://doi.org/10.1007/s44150-022-00035-y

2. Goodfellow IJ, Pouget-Abadie J, Mirza M, Xu B, Warde-Farley D, Ozair S, Bengio Y (2014) Generative adversarial nets. In: Proceedings of the 27th international conference on neural information processing systems, Montreal, Canada, vol 2
3. Johnson-Laird PN (1989) Analogy and the exercise of creativity. In: Vosniadou S, Ortony A (eds) Similarity and analogical reasoning. Cambridge University Press, New York, pp 313–331
4. Iouguina A, Dawson JW, Hallgrimsson B, Smart G (2014) Biologically informed disciplines: a comparative analysis of bionics, biomimetics, biomimicry, and bio-inspiration among others. Int J Des Nat Ecodynam 9(3):197–205. https://doi.org/10.2495/DNE-V9-N3-197-205
5. Schön DA (2017) The reflective practitioner: how professionals think in action. Routledge, London
6. Zheng X, Hu X, Arista R, Lu J, Sorvari J, Lentes J, Ubis F, Kiritsis D (2022) A semantic-driven tradespace framework to accelerate aircraft manufacturing system design. J Intell Manuf. https://doi.org/10.1007/s10845-022-02043-7
7. Hart LE (2015) Introduction to model-based system engineering (MBSE) and SYSML. In: Delaware Valley INCOSE chapter meeting. Ramblewood Country Club Mount Laurel, New Jersey. https://www.incose.org/docs/default-source/delaware-valley/mbse-overview-incose-30-july-2015.pdf
8. Yang P, Lu J, Feng L, Wu S, Wang G, Kiritsis D (2021) A knowledge management approach supporting model-based systems engineering. In: World conference on information systems and technologies. Springer, pp 581–590. https://doi.org/10.1007/978-3-030-72651-5_55
9. Mas F, Rios J, Menendez JL et al (2013b) A process-oriented approach to modeling the conceptual design of aircraft assembly lines. Int J Adv Manuf Technol 67(1–4):771–784. https://doi.org/10.1007/s00170-012-4521-5
10. Arista R, Mas F, Oliva M, Morales-Palma D (2019) Applied ontologies for assembly system design and management within the aerospace industry. In: Joint Ontology Workshops (JOWO) 2019 in Graz, CEUR workshop proceedings. http://ceur-ws.org/Vol-2518/paper-FOMI1.pdf
11. Arista R, Mas F, Vallellano C (2020) Initial approach to an industrial resources ontology in aerospace assembly lines. In: IFIP International conference on product lifecycle management. Springer, pp 285–294. https://doi.org/10.1007/978-3-030-62807-9_23
12. Mas F, Racero J, Oliva M, Morales-Palma D (2018) A preliminary methodological approach to models for manufacturing (MFM). In: IFIP international conference on product lifecycle management. Springer, pp 273–283. https://doi.org/10.1007/978-3-030-01614-2_25
13. Naqvi N, Islam MS, Iqbal M, Kanwal S, Khan A, Ye Z (2022) Deep neural combinational model (DNCM): digital image descriptor for child's independent learning. Multimed Tools Appl. https://doi.org/10.1007/s11042-022-12291-7
14. Degadwala S, Vyas D, Biswas H, Chakraborty U, Saha S (2021) Image captioning using inception V3 transfer learning model. In: 2021 6th International Conference on Communication and Electronics Systems (ICCES). IEEE, pp 1103–1108
15. Minoofam SAH, Bastanfard A, Keyvanpour MR (2021) TRCLA: a transfer learning approach to reduce negative transfer for cellular learning automata. In: IEEE transactions on neural networks and learning systems. IEEE. https://doi.org/10.1109/TNNLS.2021.3106705
16. Kuznetsova P, Ordonez V, Berg TL, Choi Y (2014) Treetalk: composition and compression of trees for image descriptions. Trans Assoc Comput Linguist 2:351–362
17. Gong Y, Wang L, Hodosh M, Hockenmaier J, Lazebnik S (2014) Improving image-sentence embeddings using large, weakly annotated photo collections. In: European conference on computer vision. Springer, Cham, pp 529–545
18. Chen L, Zhang H, Xiao J, Nie L, Shao J, Liu W, Chua TS (2017) Sca-CNN: spatial and channel-wise attention in convolutional networks for image captioning. In: Proceedings of the IEEE conference on computer vision and pattern recognition, pp 5659–5667

19. Fu K, Jin J, Cui R, Sha F, Zhang C (2017) Aligning where to see and what to tell: image captioning with region-based attention and scene-specific contexts. IEEE Trans Pattern Anal Mach Intell 39(12):2321–2334
20. Mao J, Xu W, Yang Y, Wang J, Huang Z, Yuille A Deep captioning with multimodal recurrent neural networks (m-rnn). arXiv preprint arXiv:1412.6632 (2014)
21. Chen, J., Dong, W., Li, M.: Image caption generator based on deep neural networks.
22. Lemley J, Bazrafkan S, Corcoran P (2017) Smart augmentation learning an optimal data augmentation strategy. IEEE Access 5:5858–5869
23. Colosimo BM, Pagani L, Grasso M (2022) Modeling spatial point processes in video-imaging via Ripley's K-function: an application to spatter analysis in additive manufacturing. J Intell Manuf. https://doi.org/10.1007/s10845-022-02055-3
24. Megahed FM, Woodall WH, Camelio JA (2011) A review and perspective on control charting with image data. J Qual Technol 43(2):83–98
25. Megahed FM, Wells LJ, Camelio JA, Woodall WH (2012) A spatiotemporal method for the monitoring of image data. Qual Reliab Eng Int 28(8):967–980
26. Qiu P (2005) Image processing and jump regression analysis, vol 599. Wiley
27. Yan H, Paynabar K, Shi J (2017) Anomaly detection in images with smooth background via smooth-sparse decomposition. Technometrics 59(1):102–114
28. Ramsay JO (2004) Functional data analysis. Encyclopedia of Statistical Sciences
29. Colosimo BM (2018) Modeling and monitoring methods for spatial and image data. Qual Eng 30(1):94–111
30. Colosimo BM, Huang Q, Dasgupta T, Tsung F (2018) Opportunities and challenges of quality engineering for additive manufacturing. J Qual Technol 50(3):233–252
31. Grasso M, Colosimo BM (2017) Process defects and in situ monitoring methods in metal powder bed fusion: a review. Measure Sci Technol 28(4):044005
32. Grasso M, Remani A, Dickins A, Colosimo B, Leach R (2021) In-situ measurement and monitoring methods for metal powder bed fusion: an updated review. Measure Sci Technol 32(11):112001
33. McCann R, Obeidi MA, Hughes C, McCarthy É, Egan DS, Vijayaraghavan RK (2021) In-situ sensing, process monitoring and machine control in laser powder bed fusion: a review. Addit Manuf 45:102058
34. Repossini G, Laguzza V, Grasso M, Colosimo BM (2017) On the use of spatter signature for in-situ monitoring of laser powder bed fusion. Addit Manuf 16:35–48
35. Andani MT, Dehghani R, Karamooz-Ravari MR, Mirzaeifar R, Ni J (2018) A study on the effect of energy input on spatter particles creation during selective laser melting process. Addit Manuf 20:33–43
36. Nassar AR, Gundermann MA, Reutzel EW, Guerrier P, Krane MH, Weldon MJ (2019) Formation processes for large ejecta and interactions with melt pool formation in powder bed fusion additive manufacturing. Sci Rep 9(1):1–11
37. Zhang Y, Hong GS, Ye D, Zhu K, Fuh JY (2018) Extraction and evaluation of melt pool, plume and spatter information for powder-bed fusion am process monitoring. Mater Des 156:458–469
38. Ly S, Rubenchik AM, Khairallah SA, Guss G, Matthews MJ (2017) Metal vapor micro-jet controls material redistribution in laser powder bed fusion additive manufacturing. Sci Rep 7(1):1–12
39. Riad R, Ros F, Hajji ME, Harba R (2022) An industrial portrait background removal solution based on knowledge infusion. Appl Intell. https://doi.org/10.1007/s10489-021-03099-3
40. Bouwmans T (2014) Traditional and recent approaches in background modeling for foreground detection: an overview. Comput Sci Rev 11:31–66
41. Rother C, Kolmogorov V, Blake A (2004) Grabcut-interactive foreground extraction using iterated graph cuts. ACM Transactions on Graphics (SIGGRAPH)

42. Chen L, Papandreou G, Kokkinos I, Murphy K, Yuille AL (2018) Deeplab: semantic image segmentation with deep convolutional nets, atrous convolution, and fully connected crfs. IEEE Trans Pattern Anal Mach Intell 40(4):834–848
43. Yu C, Wang J, Peng C, Gao C, Yu G, Sang N (2018) Learning a discriminative feature network for semantic segmentation. In: 2018 IEEE/CVF conference on computer vision and pattern recognition, pp 1857–1866
44. Zhang Z, Zhang X, Peng C, Xue X, Sun J (2018) Exfuse: enhancing feature fusion for semantic segmentation. In: Proceedings of the European Conference on Computer Vision (ECCV), pp 269–284
45. Otsu N (1979) A threshold selection method from gray-level histograms. IEEE Trans Syst Man Cybern 9(1):62–66
46. Li Y, Sun J, Tang CK, Shum HY (2004) Lazy snapping. ACM Trans Graph 23(3):303–308
47. Chan TF, Vese LA (2001) Active contours without edges. IEEE Trans Image Process 10(2):266–277
48. Ronneberger O, Fischer P, Brox T (2015) U-net: convolutional networks for biomedical image segmentation. In: International conference on medical image computing and computer-assisted intervention. Springer, pp 234–241
49. Hajikhani A, Suominen A (2022) Mapping the sustainable development goals (SDGs) in science, technology and innovation: application of machine learning in SDG-oriented artefact detection. Scientometrics. https://doi.org/10.1007/s11192-022-04358-x
50. Giovannini E, Niestroy I, Nilsson M, Roure F, Spanos M (2015) The role of science, technology and innovation policies to foster the implementation of the Sustainable Development Goals (SDGs) report of the expert group " Follow-up to Rio + 20, Notably the SDGs". European Commission. https://doi.org/10.2777/485757
51. Ranaei S, Suominen A, Dedehayir O (2017) A topic model analysis of science and technology linkages: a case study in pharmaceutical industry. In: 2017 IEEE Technology and Engineering Management Society Conference, TEMSCON 2017. https://doi.org/10.1109/TEMSCON.2017.7998353
52. Yau CKCK, Porter A, Newman N, Suominen A (2014) Clustering scientific documents with topic modeling. Scientometrics 100(3):767–786. https://doi.org/10.1007/s11192-014-1321-8
53. Yoon B, Phaal R (2013) Structuring technological information for technology roadmapping: data mining approach. Technol Anal Strategic Manage 25(9):1119–1137. https://doi.org/10.1080/09537325.2013.832744
54. Mikolov T, Chen K, Corrado G, Dean J (2013) Efficient estimation of word representations in vector space. In: 1st International conference on learning representations, ICLR 2013—Workshop Track Proceedings
55. Bernabei M, Eugeni M, Gaudenzi P, Costantino F (2022) Assessment of smart transformation in the manufacturing process of aerospace components through a data-driven approach. Glob J Flexible Syst Manage. https://doi.org/10.1007/s40171-022-00328-7
56. Steinbring M, Motschnig R, Pitner T (2013) A multidimensional model of web-induced flexibility in organisations. Glob J Flexible Syst Manage 14(1):57–65
57. Kovrigin EA, Vasiliev VA (2020) Barriers in the integration of modern digital technologies in the system of quality management of enterprises of the aerospace industry. In: Proceedings of the 2020 IEEE international conference "Quality Management, Transport and Information Security, Information Technologies", IT and QM and IS 2020, pp 331–335
58. Dewey H, DeVries DR (2018) Case study in utilizing the Internet of Things as a PHM architecture for aerospace applications. In: IEEE aerospace conference. IEEE
59. Dohn NB, Kafai Y, Mørch A, Ragni M (2022) Survey: artificial intelligence, computational thinking and learning. KI-Künstliche Intell. https://doi.org/10.1007/s13218-021-00751-5
60. Russell S, Norvig P (2010) Artificial intelligence: a modern approach, 3rd edn. Prentice Hall (in English), Upper Saddle River, NJ

61. Wing J (2011) Research notebook: computational thinking—what and why? The Link Magazine. Available: https://www.cs.cmu.edu/link/research-notebook-computational-thinking-what-and-why
62. Dohn NB (2021) Computational thinking - indplacering i et landskab af it-begreber. In: Dohn NB, Mitchell R, Chongtay R (eds) Computational thinking—Teoretiske, Empiriske og Didaktiske Perspektiver. Samfundslitteratur, Frederiksberg, pp 31–60
63. Goebel R et al (2018) Explainable AI: the new 42? In: Holzinger A, Kieseberg P, Tjoa AM, Weippl E (eds) Machine learning and knowledge extraction CD-make 2018 (lecture notes in computer science, No. 11015). Springer International Publishing, Cham, pp 295–303
64. Wing J (2006) Computational thinking. Commun ACM 49(3):33–35. https://doi.org/10.1145/1118178.1118215
65. Grover S, Pea R (2013) Computational thinking in K-12: a review of the state of the field. Educ Res 42(1):38–43
66. Kafai YB, Burke Q (2014) Connected code: why children need to learn programming (MacArthur Foundation Series on Digital Media and Learning). MIT Press, Cambridge, MA
67. Kafai YB (2016) From computational thinking to computational participation in K–12 education. Commun ACM 59(8):26–27. https://doi.org/10.1145/2955114
68. Hutchins E (1995) Cognition in the wild (no, nr. 1995). MIT Press, Cambridge, MA
69. Caeli EN, Yadav A (2020) Unplugged approaches to computational thinking: a historical perspective. TechTrends 64(1):29–36. https://doi.org/10.1007/s11528-019-00410-5
70. Lu JJ, Fletcher GHL (2009) Thinking about computational thinking. SIGCSE Bullet 41(1):260–264. https://doi.org/10.1145/1539024.1508959
71. Resnick M (2014) Give P's a chance: projects, peers, passion, play. In: Proceedings of Constructionism and Creativity Conference, Vienna, Austria
72. Fischer G (1994) Domain-oriented design environments. Automated Softw Eng 1(2):177–203. https://doi.org/10.1007/BF00872289
73. Thomson L, Kamalaldin A, Sjödin D, Parida V (2021) A maturity framework for autonomous solutions in manufacturing firms: the interplay of technology, ecosystem, and business model. Int Entrep Manage J. https://doi.org/10.1007/s11365-020-00717-3
74. Porter ME, Heppelmann JE (2014) How smart, connected products are transforming competition. Harv Bus Rev
75. Parida V, Sjödin D, Reim W (2019) Reviewing literature on digitalization, business model innovation, and sustainable industry: past achievements and future promises. Sustainability (Switzerland) 11(2):391. https://doi.org/10.3390/su11020391
76. Iansati M, Lakhani KR (2014) Digital Ubiquity—how connections, sensors, and data are revolutionising business. Harv Bus Rev
77. Eisenhardt KM (1989) Building theories from case study research. Acad Manage Rev 14(4):532–550. https://doi.org/10.1016/s0140-6736(16)30010-1
78. Eisenhardt KM, Graebner ME (2007) Theory building from cases: opportunities and challenges. Acad Manage J 50(1):25–32. https://doi.org/10.1002/job
79. Glaser B, Strauss A (1967) The discovery of grounded theory. Weidenfield and Nicholson, London
80. Gupta S, Modgil S, Bhattacharyya S, Bose I (2021) Artificial intelligence for decision support systems in the field of operations research: review and future scope of research. Ann Oper Res. https://doi.org/10.1007/s10479-020-03856-6
81. Wright SA, Schultz AE (2018) The rising tide of artificial intelligence and business automation: developing an ethical framework. Bus Horiz 61(6):823–832
82. Svozil D, Kvasnicka V, Pospichal J (1997) Introduction to multi-layer feed-forward neural networks. Chemom Intel Lab Syst 39(1):43–62
83. Kouziokas GN, Perakis K (2017) Decision support system based on artificial intelligence, GIS and remote sensing for sustainable public and judicial management. Eur J Sustain Dev 6(3):397

84. Rygielski C, Wang JC, Yen DC (2002) Data mining techniques for customer relationship management. Technol Soc 24(4):483–502
85. Gunasekaran A, Patel C, Tirtiroglu E (2001) Performance measures and metrics in a supply chain environment. Int J Oper Prod Manag 21(1/2):71–87
86. Lemaignan S, Warnier M, Sisbot EA, Clodic A, Alami R (2017) Artificial cognition for social human–robot interaction: an implementation. Artif Intell 247:45–69
87. Pullan TT, Bhasi M, Madhu G (2013) Decision support tool for lean product and process development. Prod Plan Control 24(6):449–464
88. Phillips-Wren G, Ichalkaranje N (eds) (2008) Intelligent decision making: an ai-based approach, vol 97. Springer, New York
89. Phillips-Wren G, Mora M, Forgionne GA, Gupta JN (2009) An integrative evaluation framework for intelligent decision support systems. Eur J Oper Res 195(3):642–652
90. Kim GH, An SH, Kang KI (2004) Comparison of construction cost estimating models based on regression analysis, neural networks, and case-based reasoning. Build Environ 39(10):1235–1242
91. Omisore MO, Samuel OW, Atajeromavwo EJ (2017) A genetic-neuro-fuzzy inferential model for diagnosis of tuberculosis. Appl Comput Inform 13(1):27–37
92. Hadavandi E, Shavandi H, Ghanbari A (2011) An improved sales forecasting approach by the integration of genetic fuzzy systems and data clustering: case study of printed circuit board. Expert Syst Appl 38(8):9392–9399
93. Lee KC, Lee N, Lee H (2012) Multi-agent knowledge integration mechanism using particle swarm optimization. Technol Forecast Soc Change 79(3):469–484. https://doi.org/10.1016/j.techfore.2011.08.004
94. Ayesta U, Erausquin M, Ferreira E, Jacko P (2016) Optimal dynamic resource allocation to prevent defaults. Oper Res Lett 44(4):451–456. https://doi.org/10.1016/j.orl.2016.04.008
95. Hu Z-H, Sheng Z-H (2015) Disaster spread simulation and rescue time optimization in a resource network. Inform Sci 298:118–135. https://doi.org/10.1016/j.ins.2014.12.011
96. Otoiu A, Titan E, Dumitrescu R (2014) Are the variables used in building composite indicators of well-being relevant? Validating composite indexes of well-being. Ecol Indic 46:575–585. https://doi.org/10.1016/j.ecolind.2014.07.019
97. Miranda SM, Kim YM (2006) Professional versus political contexts: institutional mitigation and the transaction cost heuristic in information systems outsourcing. MIS Q:725–753
98. Mata J, De Miguel I, Duran RJ, Merayo N, Singh SK, Jukan A et al (2018) Artificial intelligence (AI) methods in optical networks: a comprehensive survey. Opt Switch Netw 28:43–57
99. Fethi MD, Pasiouras F (2010) Assessing bank efficiency and performance with operational research and artificial intelligence techniques: A survey. Eur J Oper Res 204(2):189–198
100. Simões B, De Amicis R, Barandiaran I, Posada J (2019) Cross reality to enhance worker cognition in industrial assembly operations. Int J Adv Manuf Technol 105(9):3965–3978. https://doi.org/10.1007/s00170-019-03939-0
101. Sand O, Büttner S, Paelke V, Röcker C (2016) Smart assembly–projection-based augmented reality for supporting assembly workers. In: International conference on virtual, augmented and mixed reality. Springer, pp 643–652
102. Boud AC, Haniff DJ, Baber C, Steiner S (1999) Virtual reality and augmented reality as a training tool for assembly tasks. In: 999 IEEE international conference on information visualization (Cat. No. PR00210), iv. IEEE, p 32
103. Zhang X, Genc Y, Navab N (2001) Mobile computing and industrial augmented reality for real-time data access. In: Proceedings of the 2001 8th IEEE international conference on emerging technologies and factory automation, 2001, vol 2. IEEE, pp 583–588
104. Grajewski D, Górski F, Zawadzki P, Hamrol A (2013) Application of virtual reality techniques in design of ergonomic manufacturing workplaces. Procedia Comput Sci 25:289–301
105. Hořejší P (2015) Augmented reality system for virtual training of parts assembly. Procedia Eng 100:699–706

106. Pathomaree N, Charoenseang S (2005) Augmented reality for skill transfer in assembly task. In: IEEE international workshop on robot and human interactive communication, 2005. ROMAN 2005. IEEE, pp 500–504
107. Peniche A, Diaz C, Trefftz H, Paramo G (2012) Combining virtual and augmented reality to improve the mechanical assembly training process in manufacturing. In: American conference on applied mathematics, pp 292–297
108. Schwald B, De Laval B (2003) An augmented reality system for training and assistance to maintenance in the industrial context
109. Nath AG, Udmale SS, Singh SK (2020) Role of artificial intelligence in rotor fault diagnosis: a comprehensive review. Artif Intell Rev. https://doi.org/10.1007/s10462-020-09910-w
110. Bachschmid N, Pennacchi P, Vania A (2002) Identification of multiple faults in rotor systems. J Sound Vib 254(2):327–366
111. Wei Y, Li Y, Xu M, Huang W (2019) A review of early fault diagnosis approaches and their applications in rotating machinery. Entropy 21(4):409
112. Chen X, Wang S, Qiao B, Chen Q (2018) Basic research on machinery fault diagnostics: past, present, and future trends. Front Mech Eng 13(2):264–291
113. Liu R, Yang B, Zio E, Chen X (2018) Artificial intelligence for fault diagnosis of rotating machinery: a review. Mech Syst Signal Process 108:33–47
114. Xia Z, Xia S, Wan L, Cai S (2012) Spectral regression based fault feature extraction for bearing accelerometer sensor signals. Sensors 12(10):13694–13719
115. Niu G, Han T, Yang BS, Tan ACC (2007) Multi-agent decision fusion for motor fault diagnosis. Mech Syst Signal Process 21(3):1285–1299. https://doi.org/10.1016/j.ymssp.2006.03.003
116. Janssens O, Slavkovikj V, Vervisch B, Stockman K, Loccufier M, Verstockt S, Van de Walle R, Van Hoecke S (2016) Convolutional neural network based fault detection for rotating machinery. J Sound Vib 377:331–345. https://doi.org/10.1016/j.jsv.2016.05.027
117. Lei J, Liu C, Jiang D (2019) Fault diagnosis of wind turbine based on long short-term memory networks. Renew Energy 133:422–432. https://doi.org/10.1016/j.renene.2018.10.031
118. Liao GP, Gao W, Yang GJ, Guo MF (2019) Hydroelectric generating unit fault diagnosis using 1-D convolutional neural network and gated recurrent unit in small hydro. IEEE Sens J 19(20):9352–9363. https://doi.org/10.1109/JSEN.2019.2926095
119. Lee YO, Jo J, Hwang J (2017) Application of deep neural network and generative adversarial network to industrial maintenance: a case study of induction motor fault detection. In: 2017 IEEE international conference on big data (big data), pp 3248–3253. https://doi.org/10.1109/BigData.2017.8258307
120. Darpe A, Gupta K, Chawla A (2006) Dynamics of a bowed rotor with a transverse surface crack. J Sound Vib 296(4–5):888–907
121. Wang N, Jiang D, Han T (2016) Dynamic characteristics of rotor system and rub-impact fault feature research based on casing acceleration. J Vibroeng 18(3):1525–1539
122. Tajik M, Movasagh S, Shoorehdeli MA, Yousefi I (2015b) Gas turbine shaft unbalance fault detection by using vibration data and neural networks. In: 2015 3rd RSI International Conference on Robotics and Mechatronics (ICROM). IEEE, pp 308–313
123. Moosavian A, Ahmadi H, Sakhaei B, Labbafi R (2014) Support vector machine and K-nearest neighbour for unbalanced fault detection. J Qual Maint Eng 20(1):65–75. https://doi.org/10.1108/JQME-04-2012-0016
124. Godoy WF, da Silva IN, Goedtel A, Palácios RHC, Lopes TD (2016) Application of intelligent tools to detect and classify broken rotor bars in three-phase induction motors fed by an inverter. IET Electr Power Appl 10(5):430–439. https://doi.org/10.1049/iet-epa.2015.0469
125. Li Z, Pedersen VGB, Wang K, He Y (2018) Fault classification and degradation assessment based on wavelet packet decomposition for rotary machinery. Lecture Notes Electr Eng 451:509–516. https://doi.org/10.1007/978-981-10-5768-7_54
126. Yu W, Huang S, Xiao W (2018) Fault diagnosis based on an approach combining a spectrogram and a convolutional neural network with application to awind turbine system. Energies. https://doi.org/10.3390/en11102561

127. Yongbo L, Xiaoqiang D, Fangyi W, Xianzhi W, Huangchao Y (2020) Rotating machinery fault diagnosis based on convolutional neural network and infrared thermal imaging. Chin J Aeronaut 33(2):427–438
128. Yuan Z, Zhang L, Duan L (2019) Multi-sourced monitoring fusion diagnosis for rotating machinery faults. In: Proceedings—annual reliability and maintainability symposium 2019-January(1), pp 1–7. https://doi.org/10.1109/RAMS.2019.8769018
129. Charalampous P, Kostavelis I, Kopsacheilis C, Tzovaras D (2021) Vision-based real-time monitoring of extrusion additive manufacturing processes for automatic manufacturing error detection. Int J Adv Manuf Technol. https://doi.org/10.1007/s00170-021-07419-2
130. Greif H (2022) Analogue models and universal machines. Paradigms of epistemic transparency in artificial intelligence. Minds Mach 32(1):111–133. https://doi.org/10.1007/s11023-022-09596-9
131. Black M (1962) Models and metaphors. Cornell University Press
132. Berkeley G (1709) An essay towards a new theory of vision. Rhames and Pepyat
133. Villata S, Araszkiewicz M, Ashley K, Bench-Capon T, Branting LK, Conrad JG, Wyner A (2022) Thirty years of artificial intelligence and law: the third decade. Artif Intell Law. https://doi.org/10.1007/s10506-022-09327-6
134. Boella G, Caro LD, Humphreys L, Robaldo L, Rossi P, Van Der Torre L (2016) Eunomos, a legal document and knowledge management system for the web to provide relevant, reliable and up-to-date information on the law. Artif Intell Law 24(3):245–283
135. Jones AJ, Sergot M (1992) Arguments and cases: an inevitable intertwining. Artif Intell Law 1(1):3–44
136. Jones A, Sergot M (1992) Deontic logic in the representation of law: towards a methodology. Artif Intell Law 1(1):45–64
137. Bench-Capon TJ, Coenen FP (1992) Isomorphism and legal knowledge based systems. Artif Intell Law 1(1):65–86
138. Lesmo L (2009) The Turin University parser at Evalita 2009. In: Proceedings of EVALITA 2009
139. Choi S, Lee H, Park E, Choi S (2022) Deep learning for patent landscaping using transformer and graph embedding. Technol Forecast Soc Change 175:121413
140. Nguyen TS, Nguyen LM, Tojo S, Satoh K, Shimazu A (2018) Recurrent neural network-based models for recognizing requisite and effectuation parts in legal texts. Artif Intell Law 26(2):169–199
141. Medvedeva M, Vols M, Wieling M (2020) Using machine learning to predict decisions of the European Court of Human Rights. Artif Intell Law 28(2):237–266
142. Medvedeva M, Üstün A, Xu X, Vols M, Wieling M (2021) Automatic judgement forecasting for pending applications of the European Court of Human Rights. In: Proceedings of the 5th workshop on automated semantic analysis of information in legal text
143. Devlin J, Chang MW, Lee K, Toutanova K (2019) BERT: pre-training of deep bidirectional transformers for language understanding. In: Proceedings of the 2019 conference of the 1030 North American chapter of the association for computational linguistics: human language technologies, NAACL-HLT 2019, vol 1. Association for Computational Linguistics, Minneapolis, MN, USA, pp 4171–4186
144. Liu P, Qiu X, Huang X (2016) Recurrent neural network for text classification with multi-task learning. In: Proceedings of the 25th international joint conference on artificial Intelligence, pp 2873–2879
145. Kim Y (2014) Convolutional neural networks for sentence classification. In: Proceedings of the 2014 Conference on Empirical Methods in Natural Language Processing (EMNLP). Association for Computational Linguistics, Doha, Qatar, pp 1746–1751
146. Lai S, Xu L, Liu K, Zhao J (2015) Recurrent convolutional neural networks for text classification. In: 28th AAAI conference on artificial intelligence
147. Du C, Huang L (2018) Text classification research with attention-based recurrent neural networks. Int J Comput Commun Control 13(1):50–61

148. Bahdanau D, Cho K, Bengio Y (2015) Neural machine translation by jointly learning to align and translate. In: 3rd international conference on learning representations, ICLR 2015
149. Vaswani A, Shazeer N, Parmar N, Uszkoreit J, Jones L, Gomez AN, Kaiser Ł, Polosukhin I (2017) Attention is all you need. In: Advances in neural information processing systems, vol 30
150. Li Y, Huang Y, Zhang M, Rajabion L (2019) Service selection mechanisms in the Internet of Things (IoT): a systematic and comprehensive study. Clust Comput. https://doi.org/10.1007/s10586-019-02984-4
151. Singla C, Mahajan N, Kaushal S, Verma A, Sangaiah AK (2018) Modelling and analysis of multi-objective service selection scheme in IoT-cloud environment. In: Sangaiah AK (ed) Cognitive computing for big data systems over IoT. Springer, New York, pp 63–77
152. Ghadimi N, Akbarimajd A, Shayeghi H, Abedinia O (2017) Application of a new hybrid forecast engine with feature selection algorithm in a power system. Int J Ambient Energy 40(5):494–503
153. Ghadimi N, Akbarimajd A, Shayeghi H, Abedinia O (2018) Two stage forecast engine with feature selection technique and improved meta-heuristic algorithm for electricity load forecasting. Energy 161:130–142
154. Mejri M, Azzouna NB (2017) Scalable and self-adaptive service selection method for the Internet of Things. Int J Comput Appl 167(10):43–49
155. Bouzary H, Chen FF (2018) Service optimal selection and composition in cloud manufacturing: a comprehensive survey. Int J Adv Manuf Technol 97(1–4):795–808
156. Yang L, Liu L, Fan Q (2017) A survey of user preferences oriented service selection and deployment in multi-cloud environment. In: Proceedings of the 2017 18th international conference on Parallel and Distributed Computing, Applications and Technologies (PDCAT). IEEE, pp 354–359
157. Asghari P, Rahmani AM, Javadi HHS (2018) Service composition approaches in IoT: a systematic review. J Netw Comput Appl 120:61–77
158. Sun Q-B, Liu J, Li S, Fan C-X, Sun J-J (2010) Internet of Things: summarize on concepts, architecture and key technology problem. J Beijing Univ Posts Telecommun 3(3):1–9
159. Wang Q, Lee B, Murray N, Qiao Y (2016) CS-Man: computation service management for IoT in-network processing. In: Proceedings of the 2016 27th Irish Signals and Systems Conference (ISSC). IEEE, pp 1–6
160. Qu L, Wang Y, Orgun MA (2013) Cloud service selection based on the aggregation of user feedback and quantitative performance assessment. In: Proceedings of the 2013 IEEE International Conference on Services computing (SCC). IEEE, pp 152–159
161. Wang S, Zheng Z, Sun Q, Zou H, Yang F (2011) Cloud model for service selection. In: Proceedings of the 2011 IEEE Conference on Computer Communications Workshops (INFOCOM WKSHPS). IEEE, pp 666–671
162. Garg SK, Versteeg S, Buyya R (2013) A framework for ranking of cloud computing services. Future Gener Comput Syst 29(4):1012–1023
163. Jin X, Chun S, Jung J, Lee K-H (2017) A fast and scalable approach for IoT service selection based on a physical service model. Inf Syst Front 19(6):1357–1372
164. Souri A, Rahmani AM, Navimipour NJ (2018) Formal verification approaches in the web service composition: a comprehensive analysis of the current challenges for future research. Int J Commun Syst 31(17):e3808
165. Asghari S, Navimipour NJ (2018) Nature inspired meta-heuristic algorithms for solving the service composition problem in the cloud environments. Int J Commun Syst 31(12):e3708
166. Neghabi AA, Navimipour NJ, Hosseinzadeh M, Rezaee A (2019) Nature-inspired meta-heuristic algorithms for solving the load balancing problem in the software-defined network. Int J Commun Syst 32(4):e3875
167. Saberi MK (2009) Open access journals with a view of journals covered in ISI. Inf Sci Technol 24(2):105–122

168. Saberi MK (2018) Intrapreneurship in public libraries: an exploratory and confirmatory factor analysis. Libr Philos Pract 17:1–15
169. Kanagaraju P, Nallusamy R (2018) Registry service selection based secured Internet of Things with imperative control for industrial applications. Clust Comput. https://doi.org/10.1007/s10586-017-1678-6
170. Rapti E, Karageorgos A, Houstis C, Houstis E (2017) Decentralized service discovery and selection in Internet of Things applications based on artificial potential fields. SOCA 11(1):75–86
171. Ezugwu AE, Shukla AK, Nath R, Akinyelu AA, Agushaka JO, Chiroma H, Muhuri PK (2021) Metaheuristics: a comprehensive overview and classification along with bibliometric analysis. Artif Intell Rev 54:4237–4316. https://doi.org/10.1007/s10462-020-09952-0
172. Brownlee J (2011) Clever algorithms: nature-inspired programming recipes. Jason Brownlee
173. Xing B, Gao WJ (2014) Introduction to computational intelligence. In: Innovative computational intelligence: a rough guide to 134 clever algorithms. Springer, Cham, pp 3–17
174. Shukla AK, Sharma R, Muhuri PK (2018) A review of the scopes and challenges of the modern real-time operating systems. Int J Embedded Real-Time Commun Syst 9(1):66–82
175. Amirbagheri K, Núñez-Carballosa A, Guitart-Tarrés L, Merigó JM (2019) Research on green supply chain: a bibliometric analysis. Clean Technol Environ Policy 21(1):3–22
176. Trianni A, Merigó JM, Bertoldi P (2018) Ten years of energy efficiency: a bibliometric analysis. Energ Effic 11(8):1917–1939
177. Cheng MY, Prayogo D (2014) Symbiotic organisms search: a new metaheuristic optimization algorithm. Comput Struct 139:98–112
178. Chelly Dagdia Z, Avdeyev P, Bayzid MS (2021) Biological computation and computational biology: survey, challenges, and discussion. Artif Intell Rev 54(6):4169–4235. https://doi.org/10.1007/s10462-020-09951-1
179. Medvedev P (2019) Modeling biological problems in computer science: a case study in genome assembly. Brief Bioinform 20(4):1376–1383
180. Boutillier P, Maasha M, Li X, Medina-Abarca HF, Krivine J, Feret J, Cristescu I, Forbes AG, Fontana W (2018) The kappa platform for rule-based modeling. Bioinformatics 34(13):i583–i592

Chapter 2
Computations and Methods

Introduction by the Editor

Industries offer multiple products and services along with their core product offering. The required combination of products and services varies with customers due to data modeling, computation methods and algorithms. The incorporation of an artificial intelligence-based architectural framework for computations and methods within the operating system or platform enables generalized and greedy models. Industry-adapted dynamic platforms apply data learning to make real-time contextual suggestions that eventually benefit both customers and businesses. The prerequisite to study this chapter would require strong analytical skills and the ability to understand the complex algorithms and the basic knowledge of statistical modeling. This chapter senses robot swarms, distributed computing on the Internet of Things, distributed computing for the next generation, membrane computing models, digital twin methods in manufacturing services, healthcare IoT applications, deep learning models for breast cancer, convolutional neural network models for visually impaired people and chest X-ray classification, cloud computing, Q-learning and LSTM models and information security protocols for tertiary industries.

CAMS computations and methods does optimality of hyper-parameters for neuro-fuzzy models. The hyper-parameters tuning involves fuzzy c-means (FCM), subtractive clustering (SC) and grid partitioning (GP). This chapter describes the optimal performance of thermal waste-to-energy (WTE).

CAMS proposes a solution to sensing-based swarm clustering problems and leverages multiple concurrent field computations with limited domain knowledge and evaluates the approach experimentally by means of simulations, showing that the functional aspects of swarm's form clusters that will reflect the existing environmental phenomena dynamics and principles.

In this chapter, CAMS introduces distributed ledger technology (DLT) and distributed intelligence (DI) and their combination which promotes IoT methods,

V. Natarajan (ed.), *Computational Artificial Intelligence and Methods for industries*, https://doi.org/10.1007/978-981-96-5277-8_2

cloud computing, mist-computing, distributed-ledger-technology and service-oriented-computing. Then service-oriented manufacturing systems have been improved by CAMS through adjacent-order correlation (AO-C) and non-adjacent-order correlation (NAO-C). With the growth of cloud technology, the service-oriented systems introduce platform as a service and software as a service. To overcome some challenges in the service-oriented systems, the researchers have come up with new technologies: cloud computing, edge computing and fog computing. These computing technologies especially have wide applications in healthcare systems. Later, IoT has been extended to Internet of Vehicles (IoV) which performs a secure and safety communication in Vehicle-to-Vehicle (V–V), Vehicle-to-Infrastructure (V–I), Infrastructure-to-Vehicle (I–V) and Vehicle-to-Pedestrian (V–P). It supports IoV supports a large range of safety applications namely emergency call, lane change warning, automatic breaking, overtaking warning and automatic speed control. In addition, it also supports non-safety applications like real time traffic statistics, vehicle localization, parking space availability, driving behavior analysis and stolen vehicle recovery. Further, IoT focusses on cyber security in which architecture for IoT edge layer infrastructure, called AI4SAFE-IoT, has been applied.

In the manufacturing and production environment, personalized products are a part of Industry 4.0. To reduce the operational cost of production of items or products, CAMS eases the operations such as prototyping, processing, planning, planting, and resourcing. It introduces a novel concept 'type and instance' in the production world.

In our IoT environment, Representational State Transfer (REST) plays an essential role. CAMS applies the architectural style of REST, the middleware of CAMS allows devices to be treated as resources with unique Uniform Resource Identifiers (URIs) and accessed using standard HTTP methods. Through research findings, AI algorithms can communicate with the middleware through RESTful API calls, enabling seamless integration of intelligent decision-making capabilities into IoT systems. Through the penetration of wearable sensor technology and the industry adoption of the Internet of Things (IoT), there is a growing demand for efficient methods to process sensor data generated by wearable devices. CAMS proposes a classification-based approach for sensor data processing in the context of assimilated wearable sensor technology within the IoT framework. For 5G and above communications, CAMS applies data science algorithms through Support Vector Machines (SVM) to analyze the sensor data. Further, once the detected patterns identified by CAMS helps to categorize the data through actionable insights and decision-making.

Next level of discussion on The Human-Centric Internet of Things (IoT) paradigm. It is an important activity. It emphasizes the integration of IoT technologies with human needs, preferences or choices, and behaviors to create intelligent and personalized experiences by means of CAMS. In this chapter, CAMS generalizes the concepts and industry applications of IoT for human-centric nature.

Nowadays software engineering for service robotics is becoming increasingly complex due to the need for multidisciplinary expertise, heterogeneous hardware, and uncertain operating environments. This necessitates variable software to customize robots for different customers and environments. However, managing this variability adds complexity and often relies on ad hoc practices, complicating software reuse, maintenance, and evolution. By investigating its drivers, practices, methods, and challenges through a prior art and multiple-case study involving industrial companies and a systematic literature review, the present research study brings out to enhance or improve the empirical understanding of software variability in service robotics. The two methodologies: (1) Systematic Literature Review and (2) Experience Report and Interviews are followed in practice. To service the robotics, CAMS helps to identify customer needs, hardware requirements and operating environment. CAMS follow the 3 main mechanisms to do servicing the robotics: (1) Behavior analysis (2) Component based architecture and (3) Parametrization and dynamic reconfiguration.

In this chapter, we introduce the application of genetic programming (GP) for music composition. Subsequently, we bring out other evolutionary computation (EC) methods. In the early 1990s, we explored music composition by a few researchers. The interest in applying EC to aesthetic and artistic domains, such as music composition, has grown significantly. Let us review focusing on the implementation of GP and EC for music composition, highlighting the variety of compositional tasks, algorithms, individual representations, and fitness measures employed. In aesthetic studies, there is more variation and generalization in algorithmic implementations compared to traditional GP experiments. CAMS are deemed fit for exploring computational creativity in music composition and aligned well on creating variability in music composition.

Let us have an industry ecosystem which has an Internet-based intelligent and sustainable style of manufacturing products that points out a transformative approach to modern industrial production, integrating advanced digital technologies with sustainable practices to optimize efficiency, productivity, and environmental impact. This chapter provides a general overview of recent research activities, developments, and cutting-edge technology in this rapidly evolving field, highlighting key technologies, methodologies, and challenges. We also discuss the integration of Internet of Things (IoT), artificial intelligence, big data analytics, and cyber-physical systems (CPS) into manufacturing processes to enable real-time monitoring, predictive maintenance, and autonomous decision-making.

Finally, this chapter updates the research activities of Membrane Computing (MC) models and Robot Controller (RC) design, their analytical results and challenges. Membrane computing methods, also known as natuRO, Enzymatic Numerical P Systems and XP colonies systems, are a computational model inspired by the structure and functioning of biological membranes, particularly cell membranes.

Machine Generated Summaries

Disclaimer: The summaries in this chapter were generated from Springer Nature publications using extractive AI auto-summarization: An extraction-based summarizer aims to identify the most important sentences of a text using an algorithm and uses those original sentences to create the auto-summary (unlike generative AI). As the constituted sentences are machine selected, they may not fully reflect the body of the work, so we strongly advise that the original content is read and cited. The auto generated summaries were curated by the editor to meet Springer Nature publication standards. To cite this content, please refer to the original papers.

Machine generated keywords: iot, smart, device, manufacturing, robot, web, cloud, digital twin, infrastructure, service, manufacture, twin, software, composition, functionality

A Field-Based Computing Approach to Sensing-Driven Clustering in Robot Swarms [1]

This is a machine-generated summary of:

Aguzzi, Gianluca; Audrito, Giorgio; Casadei, Roberto; Damiani, Ferruccio; Torta, Gianluca; Viroli, Mirko: A field-based computing approach to sensing-driven clustering in robot swarms [1]

Published in: Swarm Intelligence (2022)
Link to original: https://doi.org/10.1007/s11721-022-00215-y

If you want to cite the papers, please refer to the original.

For technical reasons we could not place the page where the original quote is coming from.

Abstract-Summary
"Swarm intelligence leverages collective behaviours emerging from interaction and activity of several "simple" agents to solve problems in various environments."

"We address the sensing-based swarm clustering problem, where clusters are defined based on both the values sensed from the environment and the spatial distribution of the values and the agents."

"We devise a solution to sensing-based swarm clustering leveraging multiple concurrent field computations with limited domain and evaluate the approach experimentally by means of simulations, showing that the programmed swarms form clusters that well reflect the underlying environmental phenomena dynamics."

Introduction

"We focus on sensing-based clustering [2], namely a clustering problem that considers both the spatial distribution of individuals and the environmental values sensed by these individuals (through sensors)."

"Among the many approaches to express (and reason in terms of) collective behaviour featuring inherent adaptivity we shall consider the field-based computing approach [3], for its suitability in addressing dynamic problems by fostering "controlled self-organisation"."

"We provide the following contributions: we provide a precise definition of the dynamic sensing-based mobile swarm clustering problem; we present a field-based approach to address the problem, and describe a novel configurable meta-algorithm for inducing self-organised clustering in a system of neighbouring-interacting robots; we provide a publicly available and reproducible simulation framework for evaluating the algorithm on a set of diverse environment configurations, from which we observe that our solution can identify various cluster shapes and cope with a certain degree of node mobility and changes in sensed phenomena."

Background and Motivation

"Collective and self-organising behaviour is programmed through a composition of functions operating on fields mapping a set of individual agents (rather than environment locations) to computational values."

"Such propagation of data from a source of a gradient outwards is captured by a broadcast function (generic in type parameter D): When we want to aggregate data in source agents, we use the block C (collect) instead [4]: In this signature, p is a potential field usually computed through gradient; acc is the logic that combines locally perceived data with that received from neighbours; local is the local data we want to collect at a point in space (e.g. a position); and null is the null data for the acc operation (e.g. if we collect a real value, the null value could be 0)."

Contribution

"Before formally defining the problem of Dynamic Sensing Based Swarm Clustering, we summarise the assumptions about the swarm devices and the environment in which they act."

"The field-based clustering algorithm for solving the Dynamic Sensing-Based Swarm Clustering problem will be discussed below."

"We address the problem of situation awareness and recognition, where a value distributed in space (e.g. temperature as measured by sensors) has to be monitored, by recognising compact clusters with similar values (e.g. spatial regions with a similar temperature)."

"Solutions for such networks do not extend well to dynamic contexts, such as micro-drone swarms monitoring an environment: in this scenario, mobility and proximity of communication are key and need to be handled by an algorithm that is resilient to changes in both values, network structure and placement in space."

Evaluation

"A clustering process starts when a node becomes a candidate."

"Why In this last scenario, we check how the system handles node failures during the clustering process (G.3)."

"Running the simulations of scenarios 1–5 we verified how much the clusters extracted follow the underlying temperature distribution in the static context."

"When drones move randomly, the algorithm starts to produce sub-optimal cluster divisions since the nodes do not care about the cluster found, and they continue to explore the area."

"Of all, the largest experiment in the IIoT scenario included 20 nodes, which may not be enough to properly evaluate the clustering algorithm; secondly, the area covered by the experiment was quite limited (a portion of an indoor lab); finally, most of the nodes in the IIoT experiment were generally static (representing pallets) and moved only when loaded and carried by a forklift."

Related Work

"The approach proposed in this paper can be used to dynamically cluster a swarm, e.g. to monitor an environment in a decentralised way."

"A characteristic of the clustering tasks considered in the present paper (and in general, of sensing-based methods, see the next section), is that besides the sensed data, a main source of information is the spatial distance between the agents."

"The approach presented here is different in several respects: first of all, the candidate leaders are determined by a characteristic of a sensed measure (e.g. local minimum); second, each candidate cluster head spawns an aggregate process to recruit other nodes within the cluster; finally, the other nodes can join more than one cluster, based on the similarity of their sensed values with the ones sensed by leaders."

Conclusion and Future Work

"Most specifically, we use the field-based paradigm to develop a novel configurable meta-algorithm promoting self-organised clustering in a swarm of neighbouring-interacting robots."

"In order to perform the evaluation, we have implemented our algorithms using the ScaFi Scala framework for field-based computing."

"It could be interesting to compare (or combine) the meta-algorithm against (with) automated swarm behaviour design methods like multi-agent reinforcement learning."

"We would like to evaluate our algorithms on real use cases, e.g. in smart logistics and precision agriculture scenarios, by implementing them on actual drones or robots."

Distributed Intelligence in the Internet of Things: Challenges and Opportunities [5]

This is a machine-generated summary of:

Alsboui, Tariq; Qin, Yongrui; Hill, Richard; Al-Aqrabi, Hussain: Distributed Intelligence in the Internet of Things: Challenges and Opportunities [5]
Published in: SN Computer Science (2021)
Link to original: https://doi.org/10.1007/s42979-021-00677-7

If you want to cite the papers, please refer to the original.

For technical reasons we could not place the page where the original quote is coming from.

Abstract-Summary

"Widespread adoption of smart IoT devices is accelerating research for new techniques to make IoT applications secure, scalable, energy-efficient, and capable of working in mission-critical use cases, which require an ability to function offline."

"The novel combination of distributed ledger technology (DLT) and distributed intelligence (DI) is seen as a practical route towards the decentralisation of IoT architectures."

"This paper surveys DI techniques in IoT and commences by briefly explaining the need for DI, by proposing a comprehensive taxonomy of DI in IoT. This taxonomy is then used to review existing techniques and to investigate current challenges that require careful attention and consideration."

"Based on the taxonomy, IoT DI techniques can be classified into five categories based on the factors that support distributed functionality and data acquisition: cloud-computing, mist-computing, distributed-ledger-technology, service-oriented-computing and hybrid."

"Our work assists system architects and developers to select the correct low-level communication techniques in an integrated IoT-to-DLT-to-cloud system architecture."

"Open research issues for distributed intelligence in IoT are discussed."

Introduction

"This article does not provide a survey on edge computing, however, this survey article reviews contributions that attempts to build a distributed intelligence techniques in IoT by utilizing various technologies, such as cloud-computing, and mist-computing."

"Related distributed intelligence challenges in the IoT, and categorize distributed intelligence approaches into different groups including: cloud-computing, mist-computing, distributed-ledger-technology, service-oriented-computing, and hybrid."

"Reviewing the relevant literature to distributed intelligence are fourfold: (1) to understand how the DI framework can be used on top of the IoT infrastructure to build sustainable IoT applications; (2) to define the important roles of the optimal DI framework; (3) how we can apply DI techniques from several technologies to solve real-world issues in the IoT domain; and (4) to outline current challenges and present potential future directions for research with respect to developing DI approaches and platforms to support a broader range of application scenarios."

Distributed Intelligence in the IoT

"DI is defined as a sub-discipline of artificial intelligence, which allows processing functionality to be distributed, enables collaboration between smart objects, and mediates data exchange to optimise communication for IoT applications."

"Challenges: Connected IoT devices in the coming future lead to a number of fundamental challenges, e.g., resource constraints, scalability, energy efficiency, security, privacy, offline capability, and Interoperability [6–8] as well as the massive amount of data produced by IoT. Intelligence Levels: is classified into two parts, including low-level and high-level."

"The main aim of achieving security mitigation is to preserve authentication, confidentiality, and availability to ensure the security of the users, infrastructures, data and devices of the IoT [9]."

"IoT devices must have the incompetence to send their data over the network."

State-of-the-Art Distributed Intelligence in IoT

"The authors of [10] proposed Cloud-IoT distributed intelligence architecture, using an efficiency and security data acquisition scheme."

"In the simplest form, the gateway applies functionalities and rules for monitoring the health of local nodes, execution of computationally extensive tasks, and filtering application parameters The authors in [11] recently proposed a heterogeneous five-layer mist, fog, and cloud-based architecture (IoHT) that is responsible for managing and routing (near-real-time) effectively."

"It also lacks supporting interoperability In [12], the authors described fog computing architecture in support of distributed intelligence in IoT. Fog nodes are considered as hardware and software architecture."

"Other research efforts that primarily supports high-level intelligence are proposed in [13–16] in which a nodes with advanced computational resources i.e., energy and processing power are responsible for managing and handling data."

"In these approaches, low-level intelligence is supported by enabling data to be shared among various IoT devices and perform computation to provide usefulness insight out of the data."

Hardware-Based Security Primitives for IoT

"The run phase is mainly responsible for obtaining sensor response in either dynamic (large variation) response to produce TRNG output or static (stable) response to generate PUF output."

"The architecture of the proposed PUF consists of a microcontroller, eight piezo sensors, eight 100 K resistors, and an ac voltage source each of which is responsible for performing a specific task."

"The proposed PUF should be considered a weak PUF as it is designed to have only one possible challenge-response pair (CRP)."

"The proposed design utilizes a microcontroller to read the open-circuit voltages (Voc) of a selection of solar cells and generate an associated response."

"A novel weak PUF design using thermistor temperature sensors is proposed by the authors in [17]."

"A microcontroller is used to compare readings from groups of thermistor temperature sensors to generate a weak response."

Challenges and Opportunities

"We have discussed several issues, and challenges that are important in a distributed intelligence approach."

"IOTA Tangle will tackle many of the IoT issues by enabling scalablity, efficient processing of data, security, and privacy [6, 13]."

"These are considered as important issues for providing an effective IoT privacy [7]."

"Device security is regarded as an important challenge, which provides support to successfully implement many IoT applications that can be affected by cyber-attack."

"Other related research questions that should be investigated including: How IoT devices should cooperate with each other to support low-level intelligence?"

"Mist computing can work on the edge of the network and can deal with issues related security, data processing, and access control, etc. This requires mist devices to be robust and flexible."

"It is a challenge to manage all the tasks concurrently when developing distributed intelligence approaches."

Conclusions

"Since IoT devices are generally resource-constrained, there is an assumption that they shall connect to centralised infrastructure to provide additional application functionality, facilitating the smart revolution."

"To overcome these issues, distributed intelligence and DLT paradigms are being investigated."

"Distributed intelligence is suitable for resource constraints IoT devices and describes where functionality should be invoked and where data should be processed while DLT will change the entire infrastructure of IoT to many IoT applications."

"We have studied the research efforts related to distributed intelligence to identify critical functions that DI platform/approach should support."

Evolutionary-Based Neuro-Fuzzy Modelling of Combustion Enthalpy of Municipal Solid Waste [18]

This is a machine-generated summary of:

Adeleke, Oluwatobi; Akinlabi, Stephen; Jen, Tien-Chien; Adedeji, Paul A.; Dunmade, Israel: Evolutionary-based neuro-fuzzy modelling of combustion enthalpy of municipal solid waste [18]

Published in: Neural Computing and Applications (2022)
Link to original: https://doi.org/10.1007/s00521-021-06870-2

If you want to cite the papers, please refer to the original.

For technical reasons we could not place the page where the original quote is coming from.

Abstract-Summary

"The viability of thermal waste-to-energy (WTE) plants and its optimal performance have informed intelligent predictive modelling of its significant variables critical to optimal energy recovery and plant operational planning using machine learning approach."

"The optimality of hyper-parameters is significant to accurate modelling of combustion enthalpy of waste in neuro-fuzzy models."

"The significant effect of hyper-parameters tuning of different clustering techniques, vis-à-vis fuzzy c-means (FCM), subtractive clustering (SC) and grid partitioning (GP), on the performance of the ANFIS model in its standalone and hybridized form was investigated."

“The ANFIS model was optimized with two evolutionary algorithms, namely particle swarm optimization (PSO) and genetic algorithm (GA), for predicting the lower heating value (LHV) of waste using the city of Johannesburg as a case study.”

“The optimal model for LHV prediction was selected based on minimum error criteria after testing the models’ performance using relevant statistical metrics like root mean square error (RMSE), mean absolute percentage error (MAPE), mean absolute deviation (MAD), relative mean bias error (rMBE) and coefficient of variation (RCoV).”

“This study established the significance of municipality-based LHV prediction model to enhance the efficiency of thermal WTE plants and the robustness of evolutionary-based neuro-fuzzy model for heating value prediction.”

Introduction

“A proximate-based heating value prediction model of poultry waste was developed by Qian and others [19] using multiple linear progression and best-fit regression model with average absolute error (AAE) of 5.98.”

“For Instance, the studies by Olatunji and others [20], Baghban and Ebadi [21] and Wang and others [22] developed general application MSW heating value prediction model using PSO-ANFIS, GA-ANFIS and ANN models, respectively.”

“This study therefore aims to fill this gap through the following objectives (i) developing a municipality-based model for specific application in predicting the combustion enthalpy of municipal solid waste to capture the peculiarity of waste generated locally using ANFIS model optimized with PSO or GA and (ii) investigating the significant effect of hyper-parameters tuning of three different clustering techniques vis-à-vis subtractive clustering (SC), grid partitioning (GP) and fuzzy c-means (FCM) on the performance of the ANFIS standalone, PSO-ANFIS and GA-ANFIS models.”

Materials and Methods

“In building ANFIS structure, clustering is a significant vital step that is required to group data points into a similar fuzzy cluster, assign membership functions (MF) and generate fuzzy inference system structure for the data.”

“Further to this, clustering in ANFIS modelling defines the optimal number and form of fuzzy rules.”

“The significant effect of clustering techniques on the performance of ANFIS model in approximating an output function has necessitated its careful choice in building ANFIS model structure.”

“Common ANFIS clustering techniques which are used are fuzzy c-means (FCM), grid partitioning (GP) and subtractive clustering (SC).”

"The significant effect of the choice of clustering techniques and hyper-parameters on the performance of the standalone ANFIS, PO-ANFIS and GA-ANFIS models developed was investigated."

"The effectiveness and reliability of the hybrid neuro-fuzzy models developed in this study for predicting the morphological composition of waste were evaluated using the hold-out data."

Results and Discussion

"All the PSO-ANFIS models across all clustering technique gave performance metrics results which are comparable."

"Similar to what was observed for the standalone ANFIS and PSO-ANFIS models, all the GA-ANFIS models are developed with different hyper-parameters of all clustering technique emerged with comparable performances."

"The least accurate models for standalone ANFIS, PSO-ANFIS and GA-ANFIS models were obtained with the SC-clustering technique."

"The choice of membership function type has significant effect on fuzzy set theory and the fuzzy inference system and consequently affects the performance of the GP-clustered ANFIS standalone, PSO-ANFIS and GA-ANFIS models significantly in this study."

"It was revealed that the triangular input membership function (tri-MF) gave the most accurate prediction in the GP-clustered ANFIS standalone, PSO-ANFIS and GA-ANFIS models."

"All models developed standalone ANFIS, PSO-ANFIS and GA-ANFIS across all clustering technique gave good outcomes based on the values of the statistical metrics."

Conclusions

"The significance of a municipality-based heating value prediction model which captures the peculiarities of waste generated locally to a successful WTE operations has been established in this study."

"A comparative assessment of the effect of the parameters of three clustering techniques on the performance of the standalone ANFIS, PSO-ANFIS and GA-ANFIS in predicting the lower heating value of waste was investigated."

"A range of values of the control parameters of the FCM, GP and SC clustering techniques were tested in the standalone ANFIS, PSO-ANFIS and GA-ANFIS models while the best model was selected based on minimum error criteria."

"Based on the result of this result, PSO-ANFIS model clustered with a GP technique with the lowest forecast error and highest accuracy (RMSE = 0.139, MAD =

0.064, MAPE = 2.536, rMBE = 0.071 and RCoV = 0.181) is more preferred for a more accurate and robust prediction of the heating value of LHV."

A Digital Twin-Driven Perception Method of Manufacturing Service Correlation Based on Frequent Itemsets [23]

This is a machine-generated summary of:

Xiang, Feng; Fan, Jie; Zhang, Xuerong; Zuo, Ying; Liu, Sheng: A digital twin-driven perception method of manufacturing service correlation based on frequent itemsets [23]
Published in: The International Journal of Advanced Manufacturing Technology (2022)
Link to original: https://doi.org/10.1007/s00170-022-08762-8

If you want to cite the papers, please refer to the original.
For technical reasons we could not place the page where the original quote is coming from.

Abstract-Summary
"Service correlation is a mix-order correlation, which is supposed to be defined as adjacent-order correlation (AO-C) and non-adjacent-order correlation (NAO-C)."

"The existing works mainly focus on AO-C without considering NAO-C, and constantly lead to the failure of composite service execution path (CSEP)."

"The basic model of AO-C, multi-order model of NAO-C, and its relevancy degree formula are proposed based on workflow and modular design."

"A perception method based on improved Apriori algorithm is designed and the relevant supporting data is collected by digital twin technology, so as to percept AO-C relevancy degree and calculate the relevancy degree of mix-order correlation in CSEP in the proposed AO-C and NAO-C models."

Introduction

"Based on the manufacturing service data, this paper studies the construction method of initial correlation among adjacent-order atomic-level manufacturing services and the multi-dimensional information logic description method of manufacturing services through analyzing the frequent itemsets of manufacturing services in atomic level task."

"In order to construct better service composition under global path structure availability, it is necessary to first discover the correlation of adjacent-order, and then deduce the Xth-NAO-C gradually from the perspective of AO-C. To overcome these problems, this paper proposes a digital twin (DT)-driven perception method of manufacturing service correlation based on frequent itemsets, which can establish the service correlation model and perceive service correlation by frequent itemsets to determine the support degree of services and service correlation."

"We establish a digital twin-driven manufacturing service correlation perception method based on frequent itemsets to improve the availability of manufacturing service composition."

Related Works

"The industrial community mainly focuses on the service-oriented workflow model systems and has developed several web service-oriented process description languages, such as the semantic web service description language OWL-S [24], WSMO [25, 26] and WSDL-S [27]."

"These process description languages play an important role in the corresponding service process modeling and software tool development [28]."

"Based on these description languages, service process modeling and software tools [29], it conducts research on service composition framework, modeling, and optimization, etc. Therefore, the existing research is still insufficient in service correlation."

"There is a lack of specific service correlation description and service correlation modeling method that can be used in actual MSC process."

"As a relationship diagram based on correlation and availability, MSC represents the relationships between services, and transforms the process of service composition into traversing the relational graph to find reachable paths [30]."

Mix-Order Correlation Modeling Based on Correlation, Workflow and Modular Design

"Based on the relative position of services, service correlation can be divided into adjacent-order correlation (AO-C), non-adjacent-order correlation (NAO-C), and multi-stage non-adjacent-order correlation (Xth-NAO-C)."

"AO-C and NAO-C actually are the special kind of Xth-NAO-C. Combinable relevancy degree (CoRD) can reflect the availability of MO-C. In addition, the sequence of adjacent-order service can be distinguished into post-order correlation (PostO-C) and pre-order correlation (PreO-C)."

"Solid line represents AO-C and dotted line represents NAO-C. In this paper, the MSC correlation is defined as follows: It is a kind of MSC correlation involving the existence of direct input or output of the manufacturing service, i.e., directed association edges."

"According to the definition of the AO-C, 1st-NAO-C between manufacturing services is divided into the following nine kinds of correlations."

"Of MSC, not only the AO-C and first order NAO-C but also Xth-NAO-C exist between manufacturing services."

Digital Twin-Driven Perception Method of AO-C Based on Frequent Itemsets

"The manufacturing services are matched according to the demand of manufacturing resources, and the candidate service combination is generated by scanning the MSC database and comparing the matched services."

"If the support of a candidate service composition is greater than or equal to the minimum support, the candidate service composition is a frequent itemset."

"The candidate 1-item service composition C1 and its corresponding support are obtained, and the frequent 1-item service composition L1 is obtained by comparing the support of C1 with the minimum support."

"L1 are combined to generate the candidate 2-item service composition C2."

"The MSC database is scanned again to obtain the support of C2 to compare with the minimum support, and the frequent 2-item service composition L2 is determined."

"The MSC database is scanned repeatedly to generate frequent service compositions of higher items to compare with the minsup, and then frequent service compositions are combined to generate the next-level candidate service compositions."

Case Study

"C_1 and C_2, respectively, represent the computation time difference between the proposed method and PSO and GA when the number of candidate services is 40."

"The results show that with the increasing number of candidate services, the computation time of each algorithm also increases."

"S_1 and S_2, respectively, represent the success rate difference between the proposed method and PSO, the proposed method, and GA when the number of candidate services is 40."

"Compared with GA and PSO, the method proposed in this paper has a higher success rate in solving the service composition problem of multi-order correlation."

"M1 and M2, respectively, represent the memory consumption difference between the proposed method and PSO and the proposed method and GA when the number of candidate services is 40."

"The proposed method in this paper is more suitable for the case of large number of candidate services and complex correlation."

Conclusions and Future Works

"The availability of MSC is not only related to the combination of AO-C, but also to the global structural characteristics of the non-adjacent CSEP."

"Since frequent itemsets mainly reflecting data-driven correlation, an AO-C perception method based on DT and improved Apriori algorithm are designed to perceive combinable services."

"The primary task is to build the basic model of the execution path under the features of structural availability and to evaluate the global structure availability of CSEP based on the study of AO-C and NAO-C. In addition, the research on the MSC correlation mainly focuses on the selection of services, while there is often a collaboration problem among services in manufacturing services, that is, whether the selected service can complete the current task and cooperate with other services to complete the overall manufacturing task together."

Smart healthcare IoT Applications Based on Fog Computing: Architecture, Applications and Challenges [31]

This is a machine-generated summary of:

Quy, Vu Khanh; Hau, Nguyen Van; Anh, Dang Van; Ngoc, Le Anh: Smart healthcare IoT applications based on fog computing: architecture, applications and challenges [31]

Published in: Complex & Intelligent Systems (2021)

Link to original: https://doi.org/10.1007/s40747-021-00582-9

If you want to cite the papers, please refer to the original.

For technical reasons we could not place the page where the original quote is coming from.

Abstract-Summary

"One limitation of cloud-based applications is the high service response time."

"To solve these challenges, optimal computing technologies have been proposed such as cloud computing, edge computing, and fog computing technologies."

"We present a common architectural framework based on fog computing for Internet of Health Things (Fog-IoHT) applications."

"Besides, we also indicate possible applications and challenges in integrating fog computing into IoT Healthcare applications."

"The analysis results indicated that there is huge potential for IoHT applications based on fog computing."

"We hope, this study will be an important guide for the future development of fog-based Healthcare IoT applications."

Introduction

"With the development of science came the advent of cloud computing technology."

"One challenge of cloud-based applications is high service response time [32]."

"In ambulance emergency scenarios, where related data collection such as hospital location, distance, ambulance vehicles, medical staff, etc. The decision-making will be calculated in seconds and directly affected the patient life, cloud computing is not feasible."

"A computational technology with truly real-time response times could save the life of patients."

"Fog computing brings cloud capabilities into LAN to reduce service response time."

"We present a common architectural framework for fog computing-based IoT applications."

"Besides, we also point out possible applications and challenges in integrating edge computing into IoT Healthcare applications."

"A short survey of fog computing-based healthcare IoT applications."

Computing Architectures

"A typical Cloud-based architecture consists of two layers: the Cloud layer consists of cloud servers that have robust storage and high-performance computing ability."

"Due to the computing, analysis and storage are performed on the Cloud, these devices without need the storage and computing abilities."

"One of the significant limitations of cloud computing is the high service response time."

"The difference between fog computing and edge computing is only implement performance."

"Server Overhead: this criterion reflects the average computational cost per request of each solution."

"With very low service response time, fog computing will be one of the most feasible solutions for Internet of Healthcare Things applications in emergency rescue scenarios and in some real-time patient monitoring applications such as patients with a history of heart disease, blood pressure and stroke."

Survey of Recent IoT Healthcare Applications

"For the purpose to reduce service response time, improve system performance, and energy efficiency, integrating fog computing into the IoHT applications are proposed and achieved some positive results."

"In [33], Sood and others proposed an IoT healthcare application based on fog computing to monitoring and diagnose a hypertension attack of the patient in real-time."

"Realizing the limitations of healthcare IoT applications based on the Cloud, Awaisi and others [34] proposed an efficient healthcare IoT based on fog computing to enhance efficiency and security."

"To address this challenge, Wang and others [35] proposed a healthcare IoT system based on fog computing to reduce patient data retrieval time."

"According to energy efficiency direction, Isa and others [36] proposed a fog computing-based architecture for healthcare IoT applications to saving and optimise energy consumption."

Discussion and Applications

"According to the performance improvement approach, fog-IoHT applications focus on reducing service response time, energy efficiency, and improving the accuracy of decision-making."

"According to the privacy and security approach, fog-IoHT applications focus on ensuring privacy and data security at fog servers."

"According to offloading approach, fog-IoHT applications focus on offloading for fog servers."

"It should be recalled that, although performance improvement and efficient offloading all have the same goals to reduce service response time, energy, and enhance the accuracy of decision-making, they have two different approaches."

"While performance improvement uses different solutions and methods to optimize performance criteria on a fog server, efficient offloading focuses on establishing an optimal assignment of tasks schema for fog servers."

"One of the main purposes of fog computing is to move the abilities of the cloud closer to end-users for the purpose of the reduced service response times, enhance reliability, efficiency energy consumption, bandwidth, computing costs."

Challenges and Future Research Directions

"Based on the task requirements, the IoT node based on the Q-learning algorithm will choose to offload tasks to the fog servers or to the cloud to load balancing and executed more efficiently. (2) An efficient scheduling algorithm to achieve energy efficiency, reliably manage and control fog servers in heterogeneous network environments needs to be considered."

"In [37], the authors introduced an efficient scheduling algorithm to saving energy for fog-based IoT systems."

"Beside, in [38], the authors also introduced an efficient scheduling algorithm for efficient resource utilization on fog servers."

"Intelligent scheduling algorithms based on the game theory approach are used to optimise the allocation of tasks. (3) In healthcare, the need for security on fog servers have been highlighted in recent studies due to potential unsafe issues from public network environments as well as resource sharing of servers [39, 40]."

Conclusions

"To provide computing methods for these applications, several computational technologies have been proposed including cloud computing, edge computing, and fog computing."

"A traditional healthcare application based on the cloud has some limitations such as high service response time and computing costs, more energy consumption and bandwidth."

"Fog computing-based IoHT applications are proposed."

"The survey and analysis results have proven that fog computing-based IoHT applications reduce latency, save energy, improve performance, reliability compared to cloud computing."

"Because the distance between the end-users and the database is shortened, fog computing has outstanding advantages over cloud computing and especially suitable for real-time IoHT applications."

"Besides, we surveyed recently the IoT applications based on fog computing in the healthcare industry."

PPSA: Privacy Preserved and Secured Architecture for Internet of Vehicles [41]

This is a machine-generated summary of:

Bevish Jinila, Y.; Merlin Sheeba, G.; Prayla Shyry, S.: PPSA: Privacy Preserved and Secured Architecture for Internet of Vehicles [41]
Published in: Wireless Personal Communications (2021)
Link to original: https://doi.org/10.1007/s11277-021-08179-8

If you want to cite the papers, please refer to the original.

For technical reasons we could not place the page where the original quote is coming from.

Abstract-Summary

"The Internet of Vehicles (IoV) is a key component of IoT. In IoV, the messages communicated between the vehicles, the infrastructure or the wearable should be reliable and has to be communicated within a short span of time."

"Existing literatures rely on layered architecture which mainly lies towards IoT architecture."

"Privacy preserved and secured four layer architecture is proposed with its protocol stack and future directions."

"By adopting the privacy preserved approach, the analysis shows that the proposed scheme PPSA provides better throughput, Packet Delivery Ratio and End-to-end delay."

Introduction

"Over fifty billion devices are expected to be connected by 2020 of which intelligent vehicles play a major part."

"This paper shows how the Internet of Vehicles (IoV) becomes a reality by connecting every vehicle to the internet."

"The continuous growth in technology has made more advances in wireless communication where each vehicle on the road is connected with other vehicles to form a network of connected cars."

"The connected vehicles become an integral component of IoV which is a mobile network which involves communication namely Vehicle to Vehicle (V-to-V), Vehicle to Infrastructure (V-to-I) and Vehicle to Wearables (V-to-W)."

"For safety applications like collision warning, pre-crash sensing and driver assistance, the position of the neighboring vehicle has to be communicated fast to the peer vehicles to avoid collision."

"The Vehicle Infrastructure Integration (VII) in United States proposes that the emergency messages should be communicated to the neighboring vehicles within half a second."

Overview of IOV

"It is a large intelligent network which consists of vehicles equipped with On Board Unit (OBU) capable of communicating with other vehicles, pedestrians and the fixed Road Side Units (RSU)."

"It includes a centralized Trusted Authority (TA), Traffic Management Cloud Server (TMCS), several Roadside Units (RSUs) and vehicles equipped with On Board Unit (OBU)."

"The OBU in a vehicle enables inter vehicle communication and the communication between itself and Road Side Units (RSU)."

"Trusted Authority (TA) The centralized TA is responsible for vehicle registration."

"Vehicles In order to use the applications of the network, each vehicle should be a registered user."

"All registered vehicles are equipped with an OBU, which can communicate with other vehicles and RSU."

"Road Side Units (RSU) The RSU are deployed across various locations and are responsible for communicating with the vehicles and other RSUs."

Challenges

"This section briefly describes about the various challenges faced by the IoV. Communication As the vehicles in the network travel at high speed, the data should be communicated faster and without drop."

"If there is a communication failure, the safety applications results are crucial."

"There is a possibility for lack of safety warning messages resulting in vehicle collisions, lack of traffic coordination resulting in heavy traffic jams which could evade the smooth transportation in the ITS."

"Precise vehicle positioning The precise positioning of the vehicle in ITS is crucial as the vehicles communicate a message based on their position."

"The vehicles have to be precisely positioned for smooth transportation."

"Security The communication between vehicles, infrastructure and wearable's has to be secured for a sustainable network."

Related Work

"Based on the involvement of heterogeneous technologies in IoV, the researchers have classified the architecture into three layers namely data acquisition layer, communication layer and analytics layer [42, 43]."

"The third layer namely the coordination computing and control layer is responsible for analyzing the data retrieved based on the applications."

"Zuowen [44] proposed a conditional privacy preserving authentication scheme for pervasive computing environment."

"Bidi and others [45] proposed a privacy preserving broadcast message authentication scheme, which employs Message Authentication Code (MAC) instead of asymmetric verification."

"An ID based safety message authentication scheme for security and trust in VANET proposed by Biswas and others [46] uses ID based scheme for signature generation and verification."

"In this scheme, priority based message verification is done based on the MAC layer priorities where emergency messages are given higher priorities."

Architectural Framework

"The environmental factors like data from neighbor vehicles, data from wearable's namely the pedestrians and sign boards are monitored by cameras or other vehicles. (b) Network coordination and communication layer (NCCL) This is the second layer which is responsible for transferring the data retrieved from sensors, RFID and cameras to the coordinating and processing layer."

"To ensure faster and efficient data communication, vehicles are clustered together and a cluster head is elected to transfer a data to a road side infrastructure when a public event is to be reported."

"Multi-signature is employed by forming a cluster of vehicles within a communication range of the infrastructure unit."

"Signature verification is another part, which should be carried out quicker for an efficient authentication system. (iii) Data processing layer (DPL) This is the third layer which is responsible for coordinating the information retrieved from the sensor through the network."

Protocols

"The IEEE 802.11p WAVE protocol integrates the functionality of physical and MAC layer and enables the communication between vehicles, the infrastructure and wearables."

"This control point integrates the heterogeneous technologies and helps in the development of IoV. To coordinate among all the vehicle nodes in a network, clusters are formed and a cluster head is elected to represent the cluster."

"In this layer, the Application Manager (AM) is responsible for handling the safety and non-safety services."

"The resource handler protocol runs on top of all the applications and handles them."

"The handler is responsible for handling services categorized in terms of safety, efficiency and business operations."

"The Safety Application Handler (SAH) handles all the safety related applications related to traffic management like traffic congestion warning, lane change warning and congestion warning."

"The Non-Safety Application Handler (NSAH) handles non-safety applications and notifies the internet based services and everyday essentials required on travel."

Security and Privacy Preserving Approach

"The cluster head verifies the validity of the message to be generated and signs on behalf of all the vehicles based on ID based authentication [47]."

"The CH will maintain the list of cluster members."

"The CH also will be assigned to a particular vehicle after the current CH leaves the cluster."

"The steps in safety message transmission from CH to cluster members are shown below."

"The CH receives the packet and verifies the validity of the cluster member."

"After creation, it broadcasts the message to all its cluster members."

"The CH waits for ACK from all its cluster members for a limited amount of Time T_w."

"Resends the encrypted message to cluster members that did not return an ACK."

Experimental Analysis

"The performance of the proposed scheme is analyzed based on metrics such as throughput, Packet Delivery Ratio (PDR) and end-to-end delay."

"Even when the cluster size is increased, the throughput remains constant. (b) Packet delivery ratio (PDR) The PDR is the ratio of number of packets sent from the source node to the number of packets received in the destination."

"The PDR is analyzed by varying the number of vehicles in the cluster."

"When the cluster size is 30, it is ideal and both the schemes have same performance. (iii) End to end delay Delay is the total latency experienced by the packet to traverse the network and move from the source to the destination."

"The delay seems to be less than 20 ms when the number of vehicles in the cluster is less than 30 and it seems to be increasing slightly."

Conclusion and Future Directions

"Privacy preserved and secured architecture is proposed to handle the vehicular applications."

"The performance of the privacy preserved architecture PPSA is compared with the 7-layered model based on metrics namely the throughput, Packet Delivery Ratio (PDR) and end-to-end delay."

"The performance analysis shows that the proposed method provides better throughput, PDR and end-to-end delay."

AI4SAFE-IoT: An AI-Powered Secure Architecture for Edge Layer of Internet of Things [48]

This is a machine-generated summary of:

HaddadPajouh, Hamed; Khayami, Raouf; Dehghantanha, Ali; Choo, Kim-Kwang Raymond; Parizi, Reza M.: AI4SAFE-IoT: an AI-powered secure architecture for edge layer of Internet of things [48]

Published in: Neural Computing and Applications (2020)
Link to original: https://doi.org/10.1007/s00521-020-04772-3

If you want to cite the papers, please refer to the original.

For technical reasons we could not place the page where the original quote is coming from.

Abstract-Summary

"Lack of an architecture for a safer IoT environment is referred to as the most important barrier in developing a secure IoT system."

"We propose a secure architecture for IoT edge layer infrastructure, called AI4SAFE-IoT. This architecture is built upon AI-powered security modules at the edge layer for protecting IoT infrastructure."

"In the proposed architecture, we define each security module and show its functionality against different threats in real-world applications."

"Due to the integration of AI security modules in a different layer of AI4SAFE-IoT, each threat in the edge layer will be handled by its corresponding security module delivered by a service."

"We compared the proposed architecture with the existing models and discussed our architecture independence of the underlying IoT layer and its comparatively low overhead according to delivering security as service for the edge layer of IoT architecture instead of embed implementation."

"We evaluated our proposed architecture based on the IoT service management score."

"The proposed architecture obtained 84.7 out of 100 which is the highest score among peer IoT edge layer security architectures."

Introduction

"Edge layer or device level plays a vital role in the IoT environment from an end-user perspective [49]."

"In the edge layer, due to the predictable location of the connected devices in the IoT infrastructures and the limitation in computational resources, the user faces a myriad of threats."

"Since there is a wide range of threats exist in each layer of IoT architecture that has interaction with edge-devices, each layer needs a dynamic secure mechanism to tackle these threats."

"Up to this time, several AI-based methods are proposed for cyber threat hunting in the edge layer of the IoT environment and obtained promising results in dealing with unseen threats [50, 51]."

"The main contributions of this paper include: Propose a secure architecture for the edge layer of IoT environments based on multiple AI-powered components."

"Design AI-engines applied in defense components based on the service-oriented architecture for securing the edge layer of IoT environments."

Related Work

"Most of the proposed architectures are only able to address specific security challenges between edge layer devices and one of the working layer of IoT architecture, following a heuristic method."

"The proposed SOA provides data privacy and confidentiality between edge layer devices and the application layer of IoT architecture."

"Vučinić and others [52] proposed object security architecture for IoT end-to-end devices in the edge layer of the smart grid domain."

"For the edge layer security, several architectures have been proposed which rely on a specific installed security mechanism on the device rather than providing a higher aspect of architecture."

"Pinto and others [53] proposed an architecture called IIoTEED for ARM-based devices in the edge layer of the IoT environment."

"Their architecture takes advantage of virtualization and trusted computing for providing a secure environment for operating system edge layer devices."

Proposed Architecture

"CTA module in the application layer has the responsibility of attributing the malicious activity that resides on edge layer devices by a profile matching engine to its original malicious actor and recommends the optimum decision (course of the action) against the threat/attack—see Algorithm 3."

"For having a security mechanism in this layer similar to typical TCP/IP protocols, following modules are provided in AI4SAFE-IoT: Network-based Firewall (FW) Since network firewalls are primitive defense mechanisms in network-based structure, at the network layer of the IoT architecture providing a rule-based mechanism for blocking suspicious traffic is essential."

"With that in mind, the proposed AI-powered modules in the network and edge layer of the proposed architecture build profiles from devices' behavior and will be able to detect, prevent and attribute malicious behaviors."

"There are two approaches that exist to implement a security module in edge layer of IoT. The first approach takes service from the application layer of architecture, which means implementing the AI engine on top of the architecture."

Results and Discussion

"Since the proposed architecture uses AI engines to detect, locate and attribute threats in a different layer of IoT environment that interacts with edge layer devices, it will be able to tackle unseen threats based on matching their profile of behaviors with each trained engine."

"It is defined by a set of security modules that have interoperability functionality in each layer of IoT architecture that interacts with edge devices."

"The proposed metric can be used to evaluate all the IoT security architectures that include the basic service delivery requirement in each layer like AI4SAFE."

"As it shows in the overview of proposed healthcare system architecture, we defined each module in the IoT base architecture as follows: Application layer The proposed healthcare system architecture provides a management server for handling requests from edge layer device and end-user request, respectively."

Conclusion

"Several security models and architectures for the edge layer of IoT have been proposed in the literature."

"Although the existing architectures are able to address some security challenges in the IoT environment, most of such architectures suffer from the two main shortcomings."

"We proposed an AI-powered architecture for IoT environments, named AI4SAFE-IoT. In AI4SAFE-IoT, we defined a three-layer architecture with security modules based on AI engines for each layer of IoT which has interaction with the edge layer devices."

"One of the main advantages of AI4SAFE-IoT architecture is its independence of operation from end-point device resources by delivering a pervasive security mechanism based on AI through service."

Deep Learning Based Breast Cancer Detection and Classification Using Fuzzy Merging Techniques [54]

This is a machine-generated summary of:

Krithiga, R.; Geetha, P.: Deep learning based breast cancer detection and classification using fuzzy merging techniques [54]
Published in: Machine Vision and Applications (2020)
Link to original: https://doi.org/10.1007/s00138-020-01122-0
Copyright of the summarized publication:

If you want to cite the papers, please refer to the original.
For technical reasons we could not place the page where the original quote is coming from.

Abstract-Summary
"Automated nuclei detection is problematic in unevenly shaped, overlapping and touching nuclei."

"It is, consequently, essential to detect single and overlapping nuclei and distinguish them from single ones for a reasonable quantitative analysis."

"Diagnosis is improved by introducing a computer-aided diagnosis system to automatically detect breast cancer tissue nuclei from whole slide images of hematoxylin and eosin stains."

"We propose a method for the automatic cell nuclei detection, segmentation, and classification of breast cancer using a deep convolutional neural network (Deep-CNN) approach."

"The main contribution of this work is the detection of nuclei using anisotropic diffusion in a filter and applying a novel multilevel saliency nuclei detection model in ductal carcinoma of breast cancer tissue."

"The detected nuclei are classified into benign and malignant cells by applying the new Deep-CNN model."

"The novel multilevel saliency nuclei detection technique is integrated with the Deep-CNN to produce an nMSDeep-CNN model that turns out to be the most accurate results with very less computation time."

Introduction

"Despite advancements in medicine, histopathological image analysis remains the most widely used method of breast cancer diagnosis."

"H&E-stained slides are mainly used in breast cancer analysis."

"The overlapping segmentation of cells in cancer cell images further complicates the process, all of which create problems for pathologists."

"To overcome these difficulties, our work concentrates on detecting breast cancer, based on signal and overlapping nuclei, and these nuclei regions are classified into benign and malignant cells."

"One of the main diagnostic studies of breast cancer is performed using convolutional neural network."

"A novel method to detect and classify cell nuclei from histopathology images is proposed."

"A design of Deep-CNN classification model to extract features from the segmented image and integration of the nMSND and Deep-CNN to produce the novel multilevel saliency deep-CNN (nMSDeep-CNN) method that classifies benign and malignant cell images much more accurately."

Related Work

"Cell detection and segmentation are usually based on the features of the nucleus, which are distinct in different types of cells."

"Sarkar and others [55] introduced a novel saliency-based sparse coding and dictionary learning method to compute image similarity between a pair of images."

"Chen and others [56] introduced a deep contour-aware network model under a unified multitask learning framework for more accurate detection and segmentation with end-to-end fully convolutional networks."

"Zheng and others [57] proposed a hybrid of the K-means and support vector machine (K-SVM) algorithms to extract useful features and classify nuclei based on morphological feature in the H&E-stained images."

"Convolutional neural networks (CNNs) have made breakthroughs in computer vision possible, specifically in image classification, achieving promising results that outperform the best traditional machine learning methods [58]."

Background

"The divergence is an operator, which takes in the vector-valued function to define this vector field and outputs a scalar-valued function measuring the change in density of the image at each point."

"The heat equation is thus modified to:where c is the function which controls the rate of diffusion at any point in the image."

"Anisotropic diffusion can be used to remove noise from the digital images without blurring edges with a constant diffusion coefficient."

"The edges can be used while removing noise from the images [59]."

Methodology

"Global color seed point is obtained by, Partitioning the preprocessed image into superpixels."

"Inspired by Wang and others [60], we use texture prior, which is a global prior to construct the saliency point, and utilize RGB color space of H&E-stained images for segmenting the breast cancer tissue."

"The final global color saliency point is accomplished by calculating the following formula: The saliency values of normal pixels in color and texture saliency point tend to be close to zero, and the high saliency value belongs to the abnormal tissue."

"For the purpose of enhancing the abnormal regions, these two saliency points that utilize different pieces of information are clustered by the following strategy. Where x and y denote the minimum of error evaluation metric which are obtained by grid search method."

"A fuzzy-based statistical region-merging (FSRM) algorithm for segmenting the detected area of H&E-stained images is adopted."

Deep-CNN Model for Feature Extraction

"The Deep-CNN comprises different, alternative layers of trainable filters and neighborhood pooling operations that derive simple to complex features from input image patches."

"Despite the plethora of deep convolutional neural networks with similar architecture, the layers in the proposed model are appropriately modified."

"The Deep-CNN model designed consists of six convolutional layers, six pooling layers, one flattening vector, two fully connected layers, and two classes."

"A convolutional layer is used to learn features from images."

"The advantages of the CNN model are local perception and weight sharing, which are fully reflected in the convolutional layer."

"A feature map shares a convolutional kernel, which is convolved with a number of feature maps from the previous layer."

"The feature map of the convolutional layer is down sampled by taking the regional maximum or average value."

"The output of the first layer acts as input to the second layer of the Deep-CNN model."

Experimental Results and Discussion

"The performance of the proposed algorithm is evaluated based on the following measures: accuracy, Mathews correlation coefficient (MCC), F1-score, precision, sensitivity, and specificity."

"It is a measure of predicted class that is correctly detected: For cancer diagnosis, a good recall value is more important than the precision."

"Our findings showed that the proposed nMSDeep-CNN model performs better than other models, with an average F1-score of 0.963 and an average IOU of 0.861."

"The principal advantages of the proposed system are: (i) The model is fully automatic and works independently, (ii) System accuracy is high when compared to other state-of-the-art methods, (iii) Our proposed method has achieved better in overlapping cell segmentation with high precision value, (iv) Computation time is very fast and the prediction rates show that the model outperforms others, and (v) The proposed method offers a considerably good performance across our datasets."

Conclusion

"The classification framework in H&E-stained breast cancer images classifies cancer into two classes of lesions, benign and malignant."

"We have designed a screening process to detect nuclei using a novel multilevel saliency nuclei detection algorithm that offers efficient results."

"It is hoped that it can be used as an efficient framework to develop diagnostic tools that assist clinicians and pathologists in the rapid evaluation of the tumorous lesions screened."

"Further, it is used for other types of stained images as well."

Digital Twin-Based Cyber Physical Production System Architectural Framework for Personalized Production [61]

This is a machine-generated summary of:

Park, Kyu Tae; Lee, Jehun; Kim, Hyun-Jung; Noh, Sang Do: Digital twin-based cyber physical production system architectural framework for personalized production [61]

Published in: The International Journal of Advanced Manufacturing Technology (2019)

Link to original: https://doi.org/10.1007/s00170-019-04653-7

If you want to cite the papers, please refer to the original.

For technical reasons we could not place the page where the original quote is coming from.

Abstract-Summary

"This paper proposes a digital twin-based cyber physical production system (CPPS) architectural framework that overcomes the performance hurdle."

"The proposed architectural framework comprises five services that are solutions to the performance hurdle of personalized production, and it operates on information based on the proposed product, process, plan, plant, and resource (P4R) information model."

"This information model is manufacturing abstraction for personalized production and is presented on a detailed level with object-orientation and the "type and instance" concept."

"The application of the digital twin-based CPPS to a micro smart factory (MSF) provides an advanced solution for the personalized production of various products."

Introduction

"Systems, a cyber physical production system (CPPS) is a physical and engineering system that can overcome the above limitations of the personalized production process, as it is designed to improve the performance of manufacturing systems [62–64]."

"The digital twin synchronizes the information and functions of the manufacturing site and is utilized for monitoring, production planning, and process control; this helps facilitate production process performance improvement [65–70]."

"If this problem is overcome and an improved production process performance is achieved, cost is reduced, and product delivery to customers becomes affordable; hence, more efficient personalized production is achieved."

"Because a combination of multiple engineering applications corresponds to an automation service for the overall production stage, a CPPS suitable for personalized production is designed from the system-of-systems (SoS) perspective, and an information model that abstracts data for use in the system is derived."

Research Background

"This section explains the key concepts of this study: CPPS, IIoT, digital twin, and service-oriented architecture (SOA)."

"Compared to the existing virtual simulation model, a digital twin (1) enables rapid modeling using a predefined configuration and functional units, (2) can take vertical integration by exchanging data through ICT, (3) can take horizontal coordination with advanced applications, and (4) can repeatedly derive performance indicators for process and systematical efficiency enhancement [65, 67, 71–75]."

"The previous works cited in this paper, a CPPS, which considers the digital twin as a central technological element, provides multiple services that can improve production efficiency [70, 71, 76–79]."

"SOA was used to coordinate various engineering applications and services based on connections."

"As SOA is designed and operated based on data created from distributed and heterogeneous elements, the internal organization provides an integrated solution [80, 81]."

Manufacturing System and Digital Twin Configuration for Personalized Production

"This section describes a micro smart factory (MSF), manufacturing system for personalized production, and digital twin configuration for achieving efficient personalized production."

"The configuration data are imported and utilized by the operation module and simulation engine that are subcomponents of the implemented digital twin application."

"The procedures digital twin creation and synchronization method are summarized as follows. (1) The required manufacturing system and the event are set using the information object generated based on the information model. (2) By utilizing the predefined manufacturing system layout in the information object, the base model and logic are imported based on metadata. (3) The plan and resource information are set to be modelled by coordinating with the imported configuration data. (4) The product and process information are input in the digital twin by coordinating with the imported logic. (5) The status information is synchronized by the operation module with vertical integration using ICT."

Digital Twin-Based CPPS for Personalized Production

"The CPPS proposed in this study provides five services: production planning, automated execution, real-time monitoring, abnormal situation notification, and dynamic response."

"The CPPS creates a production plan and controls the site based on the service composition of multiple CPPS applications in the manufacturing system, and it also recognizes situations and provides dynamic responses."

"The main functions in the procedure employed by the digital twin application constitute the following three main steps, which are executed by the operation module with configuration data in the digital twin application: I. Digital twin creation: This step replicates the configuration and functional units of the digital twin based on the P4R information model."

"The advanced scheduling application and the digital twin application create and verify schedules using the P4R information, created production plans, and features."

Case Study

"The context-aware application sets four abnormal situations that could occur in the MSF and that required dynamic reactions: production plan change, equipment failure, production plan delay, and product quality problem."

"As observed in Algorithm 2, a production plan, which was output by the advanced planning application, and a transport matrix, which was output by the digital twin application, were provided, and an initial population was created based on information using the structure of the P4R information model."

"The developed simulation-based optimization algorithm considered the time required to implement an allocated production plan (makespan) as the fitness indicator."

"The digital twin-based CPPS first provided a production planning service for the MSF operation."

"Based on the plant, process, and resource data, the advanced planning application derived a production plan."

Conclusion

"A digital twin-based CPPS scenario and framework for handling the performance hurdle of personalized production were proposed."

"The application and implementation of the digital twin-based CPPS to the MSF showed that the proposed CPPS for the personalized production could be easily applied in accordance with the given scenario."

"Based on the services proposed through a combination of the five applications of the implemented CPPS (digital twin application, context-aware application, advanced planning application, advanced scheduling application, and device control application), various product groups and dynamic situations in the personalized production were effectively handled."

"From the perspective of the access hurdle, the automatic extraction of customer requirements and coordination with the proposed digital twin-based CPPS should be attempted; further, it should be performed to collect accurate customer requirement and to reduce the time gap between customer interaction and the manufacturing of the personalized production."

Construction and Application of Smart Factory Digital Twin System Based on DTME [82]

This is a machine-generated summary of:

Xia, Luyao; Lu, Jianfeng; Zhang, Hao; Xu, Mengying; Li, Zhaojia: Construction and application of smart factory digital twin system based on DTME [82]
Published in: The International Journal of Advanced Manufacturing Technology (2022)
Link to original: https://doi.org/10.1007/s00170-022-08971-1

If you want to cite the papers, please refer to the original.
For technical reasons we could not place the page where the original quote is coming from.

Abstract-Summary
"The digital twin system (DTS), which has a single domain, short time cycle, and unfulfillable services, cannot fully reflect the interaction and integration of the physical and informational world required by smart manufacturing."

"Research on the smart factory DTS (SFDTS) construction and application with cross-domain, multiple models have important influence on smart manufacturing."

"It analyzes the construction requirements of the DTME for a factory DTS(FDTS), product DTS(PDTS), and supply chain DTS(SCDTS) from the perspective of the life cycle."

"The smart factory DTS architecture is applied to the digital and intelligent upgrading of the hydraulic cylinder factory."

Introduction

"The digital twin optimizes business processes in various domains, and drives the overall efficient, intelligent operations of a smart factory."

"Park and others proposed a digital twin-based CPS production system framework and applied it to the micro-smart factory to reduce production time [61]."

"Digital twins are also applied to the factory or product life cycle [83]."

"Although some scholars have considered the construction of the DTS from the factory or product life cycle perspective, the lack of multidomain digital twin systems in a smart factory to interact and coevolve across domains throughout the life cycle cannot support the overall intelligent management of a smart factory."

"The paper analyzes the virtual and real integration and cross-domain interconnection and coevolution process of digital twins in three domains in the whole life cycle and proposes the concept of the digital twin manufacturing ecosystem (DTME)."

Digital Twin Manufacturing Ecosystem

"BIM denotes an information management system for the factory life cycle, including the factory digital embryo model, factory digital mapping model, factory twin intelligent model, factory reconfiguration model, factory design planning documents, factory construction documents, and factory building information standard system."

"As a digital embryo that precedes the physical product, it is also the beginning and only model accumulating product life-cycle data and integrating the 3D geometric product model, product-related attribute information, process information, and other data."

"The product digital embryo model and design documents are passed to the manufacturing stage."

"The digital twin is the data center of the product life cycle and depicts all the information and models of the product from the design and use to the scrap/recovery stage, including the geometric characteristics, performance, status, and functions of the product."

Implementing Smart Factory Digital Twin Systems Based on Digital Twin Manufacturing Ecosystems

"In the product and factory life cycle, with virtual and real mapping and intelligent matching between the services provided by the SFDTS and customer needs, the resource and information flows in the SFDTS must be quickly and efficiently adjusted, causing the logical updating and upgrading of the production layout in the SFDTS."

"In the initial plant planning stage, the SFDTS organically combines the factory to be built with product production/logistics, realistically simulates the dynamic changes in production and logistics of the manufacturing system on a computer, and analyzes the simulation data of the production/logistics system using the system analysis method."

"In the product manufacturing stage, the product status data and real-time production data are collected through sensors and transmitted to the DTS through the production system network for data processing, storage, and analysis to realize real-time mapping between the production process in the physical factory and SFDTS."

Case Study

"By collecting and processing data and using communication technology to make the data flow efficiently between the virtual and real aspects and between the three modules, HCFDTS can completely simulate the business and production processes of the hydraulic cylinder factory."

"Aiming at the problem that considerable WIP exists in hydraulic cylinder factories and that the delivery time is prolonged, the factory layout for reasonable production and operation can be planned based on the factory production/logistics scheduling simulation results."

"The guidance information generated by production optimization simulation, production process analysis, and other functions was integrated into virtual space, and the services provided by the system for users at different levels of the factory must be merged into the factory site (i.e., the virtual information in virtual space is used to guide the physical factory to change its manufacturing mode)."

Challenge

"Due to the immature development of industrial networks and data computing technology, it is challenging to meet the needs of synchronization and establish a closed loop between physical factories and virtual space in the SFDTS."

"The research on the SFDTS must adopt system engineering ideas based on the core data/models."

"This paper presents a study of the states and methods of model and data transmission in the system only."

"The idea of "governance" must be adopted in follow-up studies on data/model sharing rules, update rules, and cross-domain update rules so that all participants can realize the cogovernance and coevolution of models and data when sharing data."

Conclusion and Future Work

"The core concept of digital twin technology is the integration of virtual and real spaces, providing a new method and ideas for developing the current manufacturing industry."

"This paper introduces the digital twin concept and analyzes its connotation and the necessity of establishing a DTME from the perspective of product manufacturing process demands."

"Future proposed research work on this topic includes the following: (1) The virtual space data in the DTS come from the physical space, and the complete and effective expression of the virtual models depends on the quantity and quality of the data."

"The complete integration and application of intelligent algorithms and twin data/models in the virtual space must be further explored."

A Framework for Model-Driven Engineering of Resilient Software-Controlled Systems [84]

This is a machine-generated summary of:

Parri, Jacopo; Patara, Fulvio; Sampietro, Samuele; Vicario, Enrico: A framework for Model-Driven Engineering of resilient software-controlled systems [84]
Published in: Computing (2020)
Link to original: https://doi.org/10.1007/s00607-020-00841-6

If you want to cite the papers, please refer to the original.
For technical reasons we could not place the page where the original quote is coming from.

Abstract-Summary
"Emergent paradigms of Industry 4.0 and Industrial Internet of Things expect cyber-physical systems to reliably provide services overcoming disruptions in operative conditions and adapting to changes in architectural and functional requirements."

"We describe a hardware/software framework supporting operation and maintenance of software-controlled systems enhancing resilience by promoting a Model-Driven Engineering (MDE) process to automatically derive structural configurations and failure models from reliability artifacts."

"We describe the framework architecture outlining roles and responsibilities in a System of Systems perspective, providing salient design traits about digital twins and data analytic agents for failure propagation modeling and analysis."

Introduction

"According to this, the main contributions of this paper are: (i) a framework supporting development of resilient cyber-physical systems through connection of digital twins representing Configuration Items with a suite of diagnostic, predictive, and prescriptive analytics based on reliability engineering artifacts; (ii) an implementation of digital twins based on the architectural pattern of reflection that facilitates creation of digital twins, adaptation to configuration changes, and verification of coherence with the actual state of the system; (iii) the description of a concrete cyber-physical system that demonstrates the results of the JARVIS project in a Smart City context and illustrates applicability and positive consequences of the framework in an MDE approach."

Related Work

"In [85], an architectural framework based on an elementary meta model for simulation analyses and online planning is presented, primarily focusing on the event-driven information flow, extracted by a cognitive system monitoring physical components, rather than on the design and configuration of adopted digital twins."

"This is the first paper proposing a framework for MDE based on a distributed and executable reflective architecture leveraging on digital twins, so as to support system resilience through runtime monitoring, remote actuation capabilities, self-adaptability and reusability of the cyber-configurations: only a few works in literature describe an architecture including some of these concepts."

"In [72], a reference model for digital twin architectures is described, exploiting Bayesian belief networks and fuzzy logic to design smart interaction controllers for cloud-based systems, focusing on operational communication modes, while neglecting reliability and resilience properties and reusable configurations."

HW/SW Architecture for Resilience

"The proposed framework implements and interprets all the responsibilities of a complete IoT architectural stack, supporting—in a holistic vision—system resilience at each tier: the perception layer enables monitoring and self-adaptation capabilities over Field System instances using smart motherboards equipped with sensors and actuators; the network/transport layer promotes a scalable message-oriented communication platform supporting the adaptation to unexpected conditions through the integration of Field Data Server and Enterprise Integration subsystems; the processing layer provides intelligence by a variety of Data Analytics Server agents capable of detecting anomalies, predicting failure events, and applying adaptive just-in-time strategies starting from the analysis of raw and refined data; finally, the application layer contributes to system resilience in two ways: while the Enterprise Information Server exploits the digital twin representation to support real-time monitoring of physical components, User Terminal and Chatbot Services keep end-users in the loop through push notifications with the dual aim of providing up-to-date information about the system status and promoting timely manual and automated reconfigurations."

A Reflective Knowledge Base for Online Monitoring and Reaction

"The Reflection architectural pattern [86] provides a mechanism to dynamically adapt, at runtime, the structure and behaviour of modeled digital twins, by splitting the domain logic in two parts: the meta level captures the types of physical components and their interconnections; the base level identifies concrete instances of physical components and their interfaces in the actual configuration of the system."

"At the base level, a DigitalTwin instance represents a concrete Configuration Item (e.g., a specific power supply unit with a serial number and a version), whose hierarchical structure enables it to become either an elementary BasicComponent or a DigitalSystem made by multiple digital twins."

"In order to enable the framework with self-adaptation and self-recovery capabilities, the Knowledge Base of the EIS is enriched with a detailed representation of IoT control commands, distinguishing among: (i) high-level commands, intended as conceptualizations about actuation and monitoring services offered by Field System instances; and, (ii) low-level commands, intended as executable operations written according to the native syntax of physical devices."

Fault Tree Analysis for Vulnerability and Resilience Assessment

"In so doing, the Data Analytics Server contributes to system resilience by facilitating development and integration of intelligent data agents able to detect failure events occurred in accordance with a set of failure modes."

"By design, the Data Analytics Server provides an agent named FaultTreeAnalyzer which reacts to detected failure events performing Fault Tree Analysis (FTA) [87] on available failure models typically expressed using the Fault Tree (FT) formalism and built by reliability experts on top of some system's configurations."

"In order to infer latent high-level failures (not directly observable through monitoring techniques or data-driven diagnosis agents), to identify which combinations of component failures lead to system failures, and to plan for system self-recovery, the FaultTreeAnalyzer agent supports, natively, a mechanism of failure propagation and Minimal Cut Sets (MCS) determination provided by the propagateFailure and computeMCS methods, implementing recursive bottom-up and top-down resolutions of the structure according to the boolean semantics of supported logical gates, respectively."

A Model-Driven Engineering Process for Resilient Systems

"The JARVIS framework promotes a correct-by-construction strategy, exploiting artifacts of the industrial practice to drive the necessary configuration of executable software instances under a Model-Driven Engineering (MDE) approach, rather than a construct-by-correction strategy where defects and errors are discovered following a code-and-fix style [88]."

"Structural artifacts may capture hierarchical system compositions and dependencies, while reliability artifacts may capture reliability requirements and system failure logic, thus implying the definition of a set of fully-disciplined configuration guidelines and the specification of ad hoc data contracts for any serving layer."

"Following the proposed MDE approach, structural aspects of system configuration are preserved by construction guaranteeing that the meta-model complies to, and is isomorphic with, the structure of the specification artifact."

"Synoptically, the MDE approach results in three relevant positive side-effects: (i) bridging the gap between the insight of SoS experts and the actual implementation of running software; (ii) reducing the probability of introducing defective system configurations; and (iii) limiting time and effort required by human operators in tuning the cyber side."

Case Study

"The LTZ access control system typically consists of one or more entrances equipped with two video surveillance cameras: the first one, named violation camera, automatically takes photos of all entering vehicles; the second one, named OCR camera, is responsible for recognising the registration numbers that are then checked with respect to authorized licenses to detect illegal accesses."

"An alternative configuration of the LTZ gate system consists of just one camera for both violation and registration number detection."

"In normal conditions the system operates in its primary configuration with both cameras properly functioning, i.e., the violation camera looking down on the gate's entrance to monitor all accesses to a restricted vehicle area, while the OCR camera pointing forward and zooming in on the license plate to automatically extract the registration number."

"In so doing, the LTZ gate system can continue to properly operate, exploiting the OCR camera for both violation and registration number detection tasks."

Conclusions

"This paper presented a hardware/software framework, designed around a SoS architecture, supporting resilience at runtime of cyber-physical systems, exploiting digital twins and failure models to improve operation, integration, maintenance, and recoverability for many application scenarios, notably including Smart City and Industrial Internet of Things contexts."

"The Knowledge Base developed around digital twins has been enriched with IoT control commands design with the aim of empowering the framework with remote actuation capabilities, enabling both recoverability and adaptability mechanisms in a proactive way: in fact, data analytic agents can perform failure detection and propagation analysis to infer disruptions and to react autonomously with resolutive actions or by activating chatbot agents for human interventions."

"Besides, the actual design of the Knowledge Base in the Enterprise Information Server lends to investigate fault/failure propagation mechanisms, i.e., how a failure of one component causes a fault of another component in the digital twin hierarchy, in so enabling various quantitative analysis and simulation techniques to drive the reconfiguration process."

Design and Development of an Indoor Navigation System Using Denoising Autoencoder Based Convolutional Neural Network for Visually Impaired People [89]

This is a machine-generated summary of:

Akilandeswari, J.; Jothi, G.; Naveenkumar, A.; Sabeenian, R. S.; Iyyanar, P.; Paramasivam, M. E.: Design and development of an indoor navigation system using denoising autoencoder based convolutional neural network for visually impaired people [89]

Published in: Multimedia Tools and Applications (2022)

Link to original: https://doi.org/10.1007/s11042-021-11287-z

If you want to cite the papers, please refer to the original.

For technical reasons we could not place the page where the original quote is coming from.

Abstract-Summary

"Identifying the current location of the users can be a difficult task for those with visual impairments."

"This work presents a detailed analysis of the recent user positioning techniques and methodologies on the indoor navigation system based on the parameters, such as techniques, cost, the feasibility of implementation, and limitations."

"This paper presents a denoising auto encoder based on the convolutional neural network (DAECNN) to identify the present location of the users."

"The proposed approach uses the de-noising autoencoder to reconstruct the noisy image and the convolution neural network (CNN) to classify the users' current position."

Introduction

"The consumer position in an indoor navigation device is difficult to locate since users with visually impaired disabilities do not know the current location."

"In [90] the authors', proposed a visual localizer, which incorporates the features of ConvNet and global optimization, to achieve robust visual localization for assisted navigation for visually impaired people."

"The visual localization method for visually impaired people's navigation is suggested, which combines the strength of de-noising autoencoder and CNN."

"In this navigation system, the current position of the users is determined using a deep learning-based classification algorithm [91, 92]."

"The key goal is to develop a method of navigation for visually disabled people in indoor environments using deep learning approaches."

"This study combines the strength of de-noising autoencoder and CNN to effectively classify the current position of the users in the navigation path."

Background and Related Work

"There is no need for sensor data from different sensors or no additional system to locate the client in the building using an image-based indoor positioning."

"The present location of the user is identified based on waves, sensors, and smartphones."

"They analyzed algorithms used for finding the user location based on the light waves."

"Kirchner and others [93] proposed the users' location identification technique based on infrared waves."

"User location is calculated based on the least square method."

"Ugar and others [94] calculated the user position by fixing the ultrasonic transmitter at known places and using the time difference arrival method to locate the user."

"José and others [95], proposed the method that combines the QR code and NFC to locate the user."

"Despite these techniques, some algorithms are utilized to find the users' position based on Received Signal Strength (RSS) and classified into three different methods such as fingerprinting, trilateration, and triangulation."

Deep Convolutional Neural Network Techniques

"Each neuron in MLP neural network takes the weighted sum of its input values."

"The first layer is an input layer, the second layer is a hidden layer consisting of several non-linear RBF activation units, and the last one corresponds to the network's final output."

"The input layer consists of an n-dimensional vector that displays to every neuron of the RBF."

"Every RBF neuron stores a prototype vector from the training set in the hidden layer."

"In the figure, it is noted that the RBF neuron is represented in a bell curve and the central value is called the center of the prototype vector or neuron."

"There are four major layer forms in CNN's architecture: the input layer, the convolutional layer, the pooling layer, and the fully connected layer."

"Fully connected layer measures and classifies class label value through a way of softmax classification."

The Denoising Auto Encoder Based Convolutional Neural Network (DAECNN)

"An autoencoder is a kind of neural network approach used for data denoising, reduce dimensionality and compress data."

"Denoising auto encoder (DAE) is applied in the proposed approach to reconstruct the original image from the corrupted input."

"The image reconstructed is then used as a model for the convolution neural network (CNN)."

"The convolution layer is often called a feature extractor layer because filters in this layer extract the image features."

"It is used to render the output image proportions as the input image."

"A fully connected layer is used to distinguish images by using a Softmax activation to connect the neurons in one layer to the neurons in another."

Experimental Analysis

"Experimental analysis, 11,190 images are collected with three different classes like Class Room1, Class Room2, and Class Room 3 in the navigation area."

"The dataset is divided into three separate datasets, the training, validation, and test datasets (80% of the images as training, 10% of the image as validation, and 10% of the images as test data)."

"The parameters for this are zoom, shear, rotation, contrast, flip, brightness, etc. and 4,94,477 images are obtained for training after the augmentation."

"To train the model, the preprocessed images are fed into the input layer."

"The following are the collected data related to navigation: 1."

"Based on the collected data, the deep learning model and mobile application has been developed."

"After completion of this application, a subject test has conducted on the developed navigation with 10 visually challenged people."

Results and Discussion

"The reconstructed image is then fed as input to the CNN classification model after the noise has been removed."

"The de-noising auto-encoder-based CNN method is used to make predictions on the test images."

"Image augmentation is applied to the dataset in order to increase the model's accuracy."

"This time the final test accuracy of the model is 87.51%."

"The model is integrated with a denoising autoencoder to remove the noise in the dataset and the CNN model has retrained with the new dataset."

"In this experimental study, we have explored the following classification approaches: Deep autoencoder (DAE), Sparse autoencoder (SAE), Multi-Layer Perceptron (MLP), CNN, Radial Basic Neural Network (RBFN), and DAECNN algorithms to identify the user location within a closed environment."

"Different assessment measures such as accuracy, precision, recall, F1-measure, and error rate are used to test the performance of classification approaches."

Conclusion and Future Work

"The DAECNN algorithm combines two different neural networks such as denoising autoencoder and convolutional neural networks to efficiently classify the users' present location."

"Denoising autoencoder is applied to reconstruct the image from the noise input which helps to improve the classification performance."

"The advantage of the DAECNN approach is (1) Reconstruct the noisy image using denoising autoencoder that helps to improve the classification performance (2) CNN model automatically extracts the prominent features without any human intervention."

Q-Learning and LSTM Based Deep Active Learning Strategy for Malware defense in Industrial IoT Applications [96]

This is a machine-generated summary of:

Khowaja, Sunder Ali; Khuwaja, Parus: Q-learning and LSTM based deep active learning strategy for malware defense in industrial IoT applications [96]

Published in: Multimedia Tools and Applications (2021)
Link to original: https://doi.org/10.1007/s11042-020-10371-0

If you want to cite the papers, please refer to the original.

For technical reasons we could not place the page where the original quote is coming from.

Abstract-Summary

"Existing studies have worked on designing complex learning algorithms or deep architectures to accurately classify malware assuming that a sufficient number of labeled examples are provided."

"In the real world, getting labeled examples is one of the major issues for training any classification algorithm."

"Recent advances have allowed researchers to use active learning strategies that are trained on a handful of labeled examples to perform the classification task, but they are based on the selection of informative instances."

"This study integrates the Q-learning characteristics into an active learning framework, which allows the network to either request or predict a label during the training process."

"We proposed the use of phase space embedding, sparse autoencoder, and LSTM with the action-value function to classify malware applications while using a handful of labeled examples."

"The experimental results show that the proposed method can achieve better accuracy than the supervised learning strategy while using few labeled requests."

"This study explores the tradeoff between classification accuracy and number of label requests via the choice of rewards and the use of decision-level fusion strategies to boost the classification performance."

Introduction

"As the sensor data is stored in the edge devices, these intermediaries are vulnerable to malware attacks and might send spoofed or fabricated information to the remote server [97]."

"A semi-supervised approach, i.e. active learning, is quite useful in domains such as autonomous driving [98, 99], malware classification [100], and cancer cell detection [101], where (a) a small number of instances are selected to train a classifier and (b) the labeling cost is reduced as only requested samples are labeled instead of the entire dataset."

"Another study [102] applies a cross-domain learning strategy, i.e. model trained on one dataset selects the most informative instance from a cross-domain dataset."

"The primary contributions of this work are as follows: An active learning framework for malware detection using PSE, SAE, and LSTM with an action-value function."

"The rest of the paper is structured as follows: Section 2 consolidates the related works for active learning and semi-supervised techniques for malware classification in IIoT scenarios."

Related Work

"To automate the process, many researchers have worked on the design of features that can effectively classify malicious applications using shallow learning techniques."

"Hassen and others [103] proposed the use of disassembled malicious binaries to extract static features to classify malware applications."

"Afonso and others [104] extracted dynamic features such as system and Android API call traces and trained them using various shallow learning methods to classify malicious applications."

"Sharmeen and others [105] performed a comparative analysis of different shallow learning methods using static, dynamic, and hybrid feature sets to classify malicious applications."

"The features are then trained with a deep learning architecture to classify malicious applications."

"Naeem and others [106] transformed the dynamic features into image visualizations and used deep learning architectures to train the model for detecting malware applications."

Methodology

"The active learning block will train the LSTM network with an action-value function using a subset of the samples on each of the transform modalities, i.e. phase space embedding derived from static features (PSE-ST), phase space embedding derived from dynamic features (PSE-DT), stacked autoencoder applied to static features (SAE-ST), and stacked autoencoder applied to dynamic features (SAE-DT), accordingly."

"It makes sense to use PSE for malware detection as the probability distribution represented by PSR from both static and dynamic features tends to change with respect to time."

"We employ another feature transformation technique, i.e. autoencoders, to extract meaningful patterns."

"The extracted features from a trained SAE are obtained for training the LSTM with an action-value function."

"The LSTM network with the action-value function is trained individually on both the PSE and SAE; therefore, the probabilities from both streams can be combined using decision-level fusion to obtain the final label."

Experimental Results

"To set a baseline, we first perform an ablation study by training the LSTM network without an action-value function using static and dynamic features."

"The same LSTM network will then be trained using PSE and SAE feature modalities to compute the classification accuracy."

"The network is then trained on the data, which includes adversarial samples to check the robustness of the LSTM model along with the PSE and SAE feature modalities, accordingly."

"We then performed another ablation study to analyze the number of requests for labels and accuracy during the training phase when using the proposed LSTM network with the action-value function."

"The results show that by using the proposed LSTM with the action-value function, the PSE-ST can achieve similar performance to that of the baselines, while only using 40% of the training data and outperforms the baseline method when trained with 50% of the data."

Hypothetical Framework for Malware Detection in the IIoT Scenario

"We present a hypothetical framework based on the proposed method to detect malware in an IIoT scenario."

"The sniffer senses the packet information over a network and stores them in the traffic database."

"The Android files stored in the traffic database are then sent to the apk extractor and provided to a sandbox environment for extracting static and dynamic features, respectively."

"For IOS, ipa binaries are emulated in the MobSF framework, whereas for Windows applications, MobSF uses a Windows Host or Windows virtual machine to perform the static analysis."

"The fetched information will then be used to update the rewards (based on the correct prediction, wrong prediction, or label request) and the model is re-trained, accordingly."

"In case, if a malware application is detected, the system admin is notified through an alert system to take necessary actions and the malware behavior signature is stored in the log database for further analysis."

Conclusion

"We proposed the use of LSTM with the action-value function as an active learner to classify malicious applications in industrial IoT settings."

"Experimental results reveal that the proposed network and PSE can achieve better results, while using a handful of the label requests, i.e. 15.12%."

"We also prove that the proposed action-value function learns through its uncertainty, while requesting or predicting a label in an active learning framework."

"We experimented to show that the proposed method is resilient to the adversarial samples and can achieve better performance than the supervised learning strategy while using only 50% of the training data, which proves the robustness of our LSTM with the action-value function."

"Another natural extension of this study is to use more powerful reinforcement learning strategies such as decomposition of action-value function [107], designing a separate target network [108], choosing a different exploration strategy [109], and so forth."

A RESTful Middleware for AI Controlled Sensors, Actuators and Smart Devices [110]

This is a machine-generated summary of:

Dipsis, Nikos; Stathis, Kostas: A RESTful middleware for AI controlled sensors, actuators and smart devices [110]
Published in: Journal of Ambient Intelligence and Humanized Computing (2019)
Link to original: https://doi.org/10.1007/s12652-019-01439-3

Abstract-Summary
"The numerous applications of internet of things (IoT) and sensor networks combined with specialized devices used in each has led to a proliferation of domain specific middleware, which in turn creates interoperability issues between the corresponding architectures and the technologies used."

"What if we wanted to use a machine learning algorithm to an IoT application so that it adapts intelligently to changes of the environment, or enable a software agent to enrich with artificial intelligence (AI) a smart home consisting of multiple and possibly incompatible technologies?"

"We answer these questions by studying a framework that explores how to simplify the incorporation of AI capabilities to existing sensor-actuator networks or IoT infrastructures making the services offered in such settings smarter."

"eVATAR+ uses JAVA server technologies enhanced by mediator functionality providing interoperability, maintainability and heterogeneity support."

Introduction

"Systematization refers to providing developers with a standard way of implementing the specific type of integrations that involve AI platforms and sensor/actuator/ smart device networks."

"We aspire to simplify the task of integrating agent AI with sensor, actuator and smart device networks and a way of achieving this is by using a familiar and easy to use API."

"Our middleware is associated with an interaction paradigm for binding AI capabilities with sensor, actuator and smart devices; the capabilities will be part of intelligent agents using different agent models [111] or architectures [112]."

"We used the notion of the avatar to conceptualize and develop a familiar, comprehensible and therefore intuitive interaction paradigm for the systematization of interactions between AI programs and heterogeneous sensor, actuator and smart device technologies."

State of the Art

"A more common approach follows a service oriented architecture (SOA) paradigm to implement middleware that provide with ways to interconnect sensor (and actuator) network nodes, IoT and smart home devices."

"None of these middleware satisfies both required characteristics for: (a) a systematic integration (providing developers with a standard way for implementing the specific type of integrations that involve AI platforms and sensors and actuator networks) and (b) transparency (abstracting the low-level details of how an AI discovers and interacts with a set of a sensors and actuators)."

"Our approach is tailor made to the particularities of integrating agent AI to sensor-actuator networks, IoT settings and smart homes offering a systematic and transparent way to achieve this (similarly to the agent based middleware) while at the same time offering a commonly used and familiar approach to API interactions."

EVATAR+

"Any existing AI software technology or IoT/sensor network middleware and gateway, sensor, actuator, sensor-actuator, smart appliance or IoT device that has the resources to make REST calls can interact with eVATAR+ and participate in the proposed architecture enabling interoperability and heterogeneity."

"Discovery in eVATAR+ is the process of finding sensors/actuators/device records in the database that are compatible to the ones described in the agent description."

"Agents and devices use their own device Ids when making GET REST calls and eVATAR+ uses these Ids to identify their mapped counterparts in the persistence layer (see below)."

"The posted JSON descriptions of agents and devices that are handled by the Jersey REST API handler (in the controller layer) of eVATAR+ are represented in the software as JPA entities (Java Persistence API—https://docs.oracle.com/javaee/6/tutorial/doc/bnbqa.html) and persisted in the database."

Case Study: Agent Capabilities in Google NEST

"Study our goal is to show how we can use eVATAR+ to enable a Jade [113] MAS agent to integrate with an Nest application in order to become an actor within a Nest smart home setting by reading sensory events within it and also creating events."

"We are going to show how the Nest application uses the eVATAR+ API to register the smart home devices and how an agent registers its interest for smart home devices by using the same API."

"Having a Nest application that interacts with the Nest devices (in the simulation) using the REST API of Nest we can pursue the goal of this case study which is to show how we can enrich an application in the Nest environment with agent capabilities."

"To achieve this, the Nest application should use the eVATAR+ API to register the simulation devices with eVATAR+."

Conclusions and Future Work

"We have presented eVATAR+, a framework with an associated middleware that binds systematically and transparently interactions between AI capabilities and existing sensor-actuator networks or IoT infrastructures, thus making the services offered in such settings smarter the context of such integrations while providing a simple and easy to use interface for developers to use."

"As part of our future work plans we will explore the possibility of providing a qualitative analysis of the followed approach and start a discussion as part of a research paper about the merits of providing systematic and transparent middleware solutions not just in the area of IoT. We will also explore the possibilities of implementing an application that integrates AI (e.g. Jade agents or use a machine learning component) with existing sensor/actuator technologies such as Google Nest, the Zigbee wireless standard [114], Arduino [115], Google Assistant [116], Amazon Alexa [117], and IFTTT [118] among others."

A Classification-Based Sensor Data Processing Method for the Internet of Things Assimilated Wearable Sensor Technology [119]

This is a machine-generated summary of:

Pradhan, Manas Ranjan; Mago, Beenu; Ateeq, Karamath: A classification-based sensor data processing method for the internet of things assimilated wearable sensor technology [119]
Published in: Cluster Computing (2022)
Link to original: https://doi.org/10.1007/s10586-022-03605-3

If you want to cite the papers, please refer to the original.
For technical reasons we could not place the page where the original quote is coming from.

Abstract-Summary
"The terahertz and large-scale processing of the 6th generation communication and computation technology is assimilated in the wearable sensor data exchange process."

"A fault-tolerant data processing method (FTDPM) is proposed to handle uneven sensor data."

"The non-uniform and unsynchronised observation interval-based sensor data is analyzed for its impact on healthcare recommendations."

"The reliability in the recommendation ensures the precise need for sensor data processing, mitigating the faults."

"This learning classifier differentiates the uniform and non-uniform sensor data traffic for providing reliable recommendations."

"The data from the uniform process is replicated in the non-uniform sequence for recommendation filling."

"This is carried out based on marginal classification and near-to-reliable data as classified using SVM."

Introduction

"The Internet of things (IoT) provides secure data acquisition and sharing through devices connected to a network in the service layer [120]."

"Limited pressure on healthcare devices and better care are provided for the high-risk patients by potential IoT solution [121]."

"Smart objects play a vital role in IoT that uses different sensors and actuators to recognize the information and build a potential network for communicating with other devices [122]."

"Big data management is required to increase IoT-based smart health monitoring systems to provide solutions to take healthcare precautions [123]."

"6G technology can provide intelligent healthcare systems that overcome 5G technology challenges by having privacy [124], security, green energy, artificial intelligence characteristics."

"Wearable monitoring systems are developed by detecting non-intrusive sensor signals and transmit them to smart devices through wireless transmission technology [125]."

Related Works

"Critical parameters are measured from the ambient domain by a proposed protocol based on patients' status, investigation, demands, etc. IoT gateway will integrate and synchronize the new wearable devices."

"Fault-tolerant mHealth framework in the context of IoT-based real-time wearable health data sensors is suggested by Albahri and others [126]."

"Sports wearable device design and health data monitoring is modeled by Zhang and others [127] based on wireless IoT. Keener physiological sensor is required to describe the development of wireless communication."

"Unsynchronised wearable sensor data analytics model is used in [128] to improve smart healthcare systems' performance."

"Time factor and data frequency are used to classify the collected wearable sensor devices."

"Zhou and others [129] recommended a wearable IoT-based cloud-assisted framework for swimming persons in the health monitoring system."

"In [130], IoT patient monitoring based on fog computing and data mining is used to use the cardiac arrhythmia case."

Proposed FTDPM

"The data with uniform and non-uniform are detected and evaluates the better recommendation on time from this processing."

"The uniform data is communicated with the IoT devices and deploys the recommendation to the response."

"The history of healthcare recommendations is used to map with the sensor data and identify the non-uniform and un-synchronization data."

"The detection of uniform, synchronized, and non-uniform un-synchronized data is done, and the below equation is used to establish the communication to improve the recommendation."

"It is a classification method that differentiates the uniform and non-uniform data from the sensor device and establishes the communication to the uniform data this SVM is used."

"The large margin selection is used to evaluate the uniform data's better processing of the recommendation system."

"This recommendation is made by deploying the FTDPM method associated with the SVM method; here, the data classification is used to provide feedback for the uniform data."

Results

"The data is sensed from the wearable device, and it is performed in the 6G environment; this is carried out to improve the recommendation system."

"The recommendation system is used to provide uniform data from wearable devices."

"This time is estimated to establish better communication among the varying data sensed from the devices."

"The detection of uniform data deploys the recommendation by performing mapping with the preceding data."

"The identification of data is made, whether it is uniform or not based on this recommendation is forwarded on time."

"The processing complexity decreases if the request holds the uniform data sensed from the wearable sensor."

"The time of responding is identified for the request from the devices that relates to the uniform data."

"The recommendation is forwarded to the requested devices, and it is associated with the differentiation of data."

Conclusion

"The proposed method utilizes the wearable sensor data for identifying the variations in different observation time intervals."

"This identification helps to prevent recommendation failures in different time intervals and non-uniform sensor data analysis."

"The processing complexity and time issues due to improper classification of non-uniform data are mitigated in this proposed method."

"The proposed method reduces processing and response time, recommendation failures, and processing complexity for different requests and communication intervals."

Effective Utilization of Multiple Convolutional Neural Networks for Chest X-Ray Classification [131]

This is a machine-generated summary of:

Rammuni Silva, Ravidu Suien; Fernando, Pumudu: Effective Utilization of Multiple Convolutional Neural Networks for Chest X-Ray Classification [131]
Published in: SN Computer Science (2022)
Link to original: https://doi.org/10.1007/s42979-022-01390-9

If you want to cite the papers, please refer to the original.
For technical reasons we could not place the page where the original quote is coming from.

Abstract-Summary
"Out of the numerous types of Medical Imaging modalities available, radiography stands out a bit more than others due to its capabilities of diagnosing diseases and conditions, including life-threatening conditions."

"Chest Radiography holds even higher importance, as it focuses a critical area of the human body."

"Interpreting a Chest Radiography image can be challenging and usually done by an experienced Radiologist for accurate results."

"This paper presents a novel way of parallelizing multiple architectures of Convolutional Neural Networks focusing on Chest X-ray classification."

"The paper further presents a comprehensive evaluation of the existing architectures with the parallelized results of them using our method."

"We used four large-scale datasets, including a non-medical one, for the evaluation of our models."

"We managed to achieve better accuracy for 9 out 13 and 11 out of 14 labels on our two main evaluation datasets."

Introduction

"The most used method in diagnosing chest-related diseases is Chest Radiography."

"As shown in [132], confusions can occur when diagnosing chest-related diseases using chest radiography due to the complex structural representations in chest radiography images, which can differ from one person to another."

"Reference [133] shows that in the United States, the percentage of active radiologists out of the whole medical professionals have been decreasing in the previous years."

"This research will also focus on a thorough re-evaluation and critical analysis of the existing works and to present them to the research community in a sensible manner."

"A critical evaluation of the existing work related to Computer-Aided Detection systems. (We suggest referring to our initial paper [134] for a more broader and critical discussion on the existing works) A requirement analysis for the proposing system via a brief survey study."

A Concise Review of Existing Work

"As per our initial paper [134] we mainly identify three components that we should consider when discussing an AI-based CXR diagnosis system."

"Image segmentation related to CXR images was done mainly in three ways; Rule-based, ML Based and DL Based."

"Reference [135] presents reliable results segmenting the lung area of CXR images nearing human-level performances."

"These identity mappings evidently work very well in fitting CXR images which have structurally very similar visual model across all the samples."

"Even though CNN architectures are capable of learning just by image data and their related labels, evidence like [136] in the literature suggests they perform better when we influence them positively to learn on the CXR images."

"Almost all the existing works merely rely on the CXR image (or more views whenever possible) to detect pathologies on CXRs."

Survey Study

"The targeted audience was the Medical Students in their final study year (majority) of MBBS degree in Sri Lanka from different state universities."

"They are the most targeted audience in the questionnaire as generally, they are the most experienced medical students."

"It could be seen that, very often, medical students get to diagnose patients by their CXR."

"One respondent (medical student) has suggested that the system might be helpful for radiologists but can be misleading for physicians."

"Interviews were explicitly done on a targeted audience to get medical-related knowledge, understanding and the requirements at a concrete level."

"The questionnaire was dedicated to the medical students to capture requirements at a higher level but from a larger audience."

"All the interviewees believe that having an automated detection report of the potential findings in a CXR would help proceed with or to confirm the interpreted results from CXR screening further."

ParallelXNet

"Those discussions show that different neural network architecture performs differently in detecting each different pathology as numerous visually different pathologies could be found using CXRs."

"For the above purpose, a commonly used technique is the Ensembling Method, which basically implies the averaged output of multiple neural networks that have been trained separately."

"This artificially provides more data for the neural network to train on, ultimately improving the model accuracy."

"It will make the neural network more biased for the learning of the label type with a higher count when training."

"Step per epoch: 4500 [Batch Size: 16] Learning Rate: 0.0001 Epochs: MIMIC-CXR Dataset-18 | ChestX-ray14 Dataset-9 In summary, for the MIMIC-CXR dataset, we trained the neural network more than 3.5 times, and for the ChestX-Ray14 dataset, 14 times over the whole dataset."

Web Application

"We selected Python [v3.7] as the primary language for the development of the model and the Server of the Web Application."

"We used a Python-based framework, Flask [v1.1.2], to develop the back end of the application."

"The Deep Learning Model, which is the most important component in the system, is placed in this tier."

"Presentation Engine: Carries out the necessary functionalities needed to present the data generated in the detection engine in an informative way onto the presentation tier."

"We used TensorFlow-based version of Keras [TensorFlow v2.4.1] to develop the deep learning model."

"Keras is an easy-to-use, high-level framework that allows developers to build, train and test ML models with a minimum number of code lines."

"Although Keras is a high-level library, its integration with TensorFlow allows the developers to do development at the low level of the library while making the best use of the high-level functionalities."

Evaluation

"We trained the existing model architectures providing the exact same conditions as in previous tests."

"Although evidently, the ParallelXNet is architecture proven to be better at accuracy, for mobile-based hardware, we recommend using DenseNet-121 due to its significantly low model size."

"The purpose of this test is to identify how the trained model generalizes between datasets."

"We note that generalization is mostly affected by the trained set of data and not by the model's architecture."

"We also note that although there are 5159 images in the test set of the MIMIC-CXR dataset, only 234 images were given as the test set in CheXpert."

"Another finding that we could derive from these test results is that the AUC score is not an absolute representation of the accuracy of the model."

"We tested our model also on the CIFAR-10 dataset to further verify the applicability and performance of it in non-medical scenarios."

Limitations and Future Work

"The built model is only trained on the 14 pathologies which are available in the datasets."

"Due to the time restrictions of the Kaggle Notebooks, continuous training was not possible, which led to training the model in parts."

"In-system model re-training: This will allow overcoming the issue of generalization. "'

"Custom model uploads: This allows users to train models and upload, making the web app universal."

"This can possibly allow even the model to consider the past CXRs of a patient when his/her new CXR are being detected through the system."

"We should note that our work shows a pathway to develop further the existing DL-based models focusing on medical-related applications."

"Although the models nor the whole system is not recommended for medical usage, our work can be further developed and trained using larger and more varied datasets to increase the accuracy and reliability."

A Scalable Hierarchically Distributed Architecture for Next-Generation Applications [137]

This is a machine-generated summary of:

Ravuri, Hemanth Kumar; Vega, Maria Torres; van der Hooft, Jeroen; Wauters, Tim; De Turck, Filip: A Scalable Hierarchically Distributed Architecture for Next-Generation Applications [137]

Published in: Journal of Network and Systems Management (2021)
Link to original: https://doi.org/10.1007/s10922-021-09618-4

If you want to cite the papers, please refer to the original.

For technical reasons we could not place the page where the original quote is coming from.

Abstract-Summary

"The rigidity of traditional network architectures, with tightly coupled control and data planes, impairs their ability to adapt to highly dynamic requirements of future application domains."

"Efforts have been made to propose alternative decentralized solutions, such as the flat distributed SDN architecture."

"When integrated seamlessly, both SDN and NFV can offer solutions to the problems posed by highly dynamic application domains."

"This work proposes a scalable hierarchical SDN control plane architecture for SDN/NFV-based next-generation application domains such as immersive media delivery system."

"We have implemented the proposed architecture based on the well-known state-of-the-art ZeroSDN controller."

"To evaluate the performance of the architecture, we have implemented an on-demand immersive media (point cloud) streaming application and varied the load on the control plane using the background traffic."

"We show that the proposed architecture performs better than the rest in terms of scalability, lost flows, and processing latency."

"Our study shows that the proposed architecture when distributed to three controllers, accepts 23% more flows with almost 70% reduced processing latency compared to the state-of-the-art ONOS controller."

Introduction

"The Software-Defined Networking (SDN) paradigm, where the control plane is decoupled from the data plane, is imperative to future Internet architectures due to its ability to assist flexible delivery and support the implementation of application-specific policies."

"Despite the flexibility offered by SDN, the conventional centralized architecture of the control plane suffers from scalability issues [138]."

"To cope with these issues, alternative SDN control plane architectures have been introduced, such as the flat and hierarchical ones [138]."

"[139], we found that the scalability of the flat distributed SDN control plane is impaired by different types of overheads inherent in the architecture."

"We aim to integrate the SDN and NFV domains in the control plane to support the next-generation applications."

"We propose a scalable hierarchical control layer architecture for next-generation application domains such as immersive multimedia delivery, that require low flow setup latency and higher scalability."

Background

"This section provides a background on SDN control plane architectures along with a detailed explanation of the flow setup process in each architecture."

"Flat SDN control plane architectures refer to the dividing of the network into multiple subdomains and assigning each of these to a controller."

"The process of handling a flow request in the centralized SDN control plane architecture is fairly simple."

"The hierarchical SDN control plane incorporates local-level event processing in the architecture."

"The local controllers subscribe to the root controller for events like processed flow requests and global network state information."

"With the separation of control and data planes, SDN enables the exchange of information between the application and network layers, allowing service providers to have an overview of the general state of the network, and to make intelligent decisions (to meet service requirements) on how to steer traffic through VNFs better [140]."

Scalable Hierarchical SDN/NFV-Based Architecture for Next-Generation Applications

"The next component in the lower tier is the module that supports local-level control functions."

"At the higher tier of the control plane are the global-event-processing modules."

"While the state-based modules can still reside on a centralized node, the policy-based modules have to be distributed to provide enough computation and process requests simultaneously."

"The local level VNF monitoring modules collectively provide the state of the infrastructure and the VNFs to the upper tier to form a global view."

"The global NFV modules can request the policy modules to route the flow through one or more intermediate nodes to provide the requested service."

"The routing module on the top tier requests the NFV modules to allocate resources and provide it with resource information to compute a path."

"The NFV Manager module provides the routing module with resource information based on the latest status update from the local monitoring modules."

Experimental Evaluation

"In order to do so, we generated ping flow requests (increased quadratically with the number of hosts per switch) along with the 14 point cloud flows, for different degrees of controller distribution."

"In the case of NSFNET, it is observed at 4 hosts per switch, while it is visible at 1 host per switch for Internet2 OS3E. In the NSFNET topology, the increase rate of the graph falls by 80% as the load increases from 2 to 6 hosts per switch (3136–28,224 flow requests) in the case of the HIER_DIS with one controller."

"As part of the experiment, we have varied the number of nodes running the routing controllet from 1 to 15, evaluating the performance in the case of NSFNET topology under the highest load condition (6 hosts per switch)."

Conclusions and Future Work

"The performance of the architecture has been evaluated in comparison to the state-of-the-art alternatives, namely, centralized and flat distributed SDN architectures for an immersive media delivery use case."

"Our hierarchically distributed approach performs better compared to the flat distributed architecture under increasing load conditions (requests and network size) when decentralized with three controller nodes."

"Our evaluation proved that the proposed architecture is more reliable (by reducing the loss), scalable, and efficient (lower flow setup latency) compared to other architectures."

"While our approach proves to have a better performance than the flat alternative, it still suffers from inherent bottlenecks as the switch adapter controller maxes out with an increasing load."

"The architecture allows the distribution of switch adapters over multiple nodes, which has been differed to future work."

Software-Defined Dew, Roof, Fog and Cloud (SD-DRFC) Framework for IoT Ecosystem: The Journey, Novel Framework Architecture, Simulation, and Use Cases [141]

This is a machine-generated summary of:

Ahammad, Ishtiaq; Khan, Md. Ashikur Rahman; Salehin, Zayed Us: Software-Defined Dew, Roof, Fog and Cloud (SD-DRFC) Framework for IoT Ecosystem: The Journey, Novel Framework Architecture, Simulation, and Use Cases [141]
Published in: SN Computer Science (2021)
Link to original: https://doi.org/10.1007/s42979-021-00521-y
Copyright of the summarized publication:

If you want to cite the papers, please refer to the original.
For technical reasons we could not place the page where the original quote is coming from.

Abstract-Summary
"We consider multi-tiered computational infrastructure which would be feasible to provide services from the nearest possible location of end devices."

"To mitigate multi-tier infrastructure issues, Software-Defined Networking (SDN) steps in."

"The journey behind SDN-supported multi-tier computational infrastructure can be understood from this paper's elaborate study."

"A novel framework termed "SD-DRFC (Software-Defined Dew, Roof, Fog and Cloud computing)" is proposed for today's IoT ecosystem."

"The role and functionality of each tier of SD-DRFC framework are adequately explained."

"A use case focused on the SD-DRFC framework is then presented and simulated by utilizing iFogSim simulator."

"To evaluate the efficiency of the presented SD-DRFC framework, four QoS parameters (Latency, Network Usage, Cost, and Energy Consumption) are considered in this paper."

"When comparing the simulation results, the presented SD-DRFC framework performs much better than cloud-only implementation."

Introduction

"End-user devices like smart devices, sensors, third-party items are connected to the computer engines or cloud/fog/roof/dew/other computing systems via the internet or intranet in a typical IoT ecosystem."

"Sensor bridge This element operates as a multi-protocol interface that processes and routes data from the local IoT network to the IoT computing services platform (e.g. cloud, fog, roof, and dew)."

"IoT services IoT gadgets or networks can be accessed utilizing IoT services that are typically deployed for universal connectivity on the computing platform, such service could be utilized for system management, and service management and any decisions can be taken based upon it."

"In this, IoT service providers offer connectivity to end-users for computing platform-based networking system management."

"RC is a decentralized IoT networking and computing platform that is always available to allow on-site real-time functionality, such as next-hop connectivity to the appropriate end-user gadgets (things or dew), real-time background creation and decision-making factors, providing ideal data routing to cloud/fog/other providers, and everything with continual protection."

Literature Survey

"As our proposed SD-DRFC framework is based on cloud, fog, roof and dew computing paradigm, survey on these four computing paradigms and their relationship with IoT ecosystem is performed here."

"Fog computing incorporates cloud to built ideal architecture for the IoT ecosystem."

"As cloud with SDN faces challenges to support the recent IoT ecosystem; hence FC is required to be merged with the combination of cloud-SDN."

"This implies that there is no framework or related research that incorporates dew, roof, fog and cloud architecture where this combined architecture is regulated by SDN."

"There is no kind of study that conducts simulation performance on the frameworks/architectures that incorporates the cloud, fog, roof and dew computing paradigms where this multi-tier architecture is managed by SDN."

"Further, there is no study or research which presents "Use Cases" based on integrated cloud, fog, roof and dew computing architecture."

Analysis on Cloud, Fog, Roof and Dew Computing Paradigms

"The Internet connectivity was expected to deliver real-time service from the mainframe phase to the post-cloud phase."

"By DC implementation, without internet access, real-time service could now be delivered."

"The Real-time Impact (RTI) metric is being used to explain the effect of internet connections on different computing system phases in the provision of real-time services."

"In the initial phase of fog, the value of G is relatively much larger than H. RTI's importance is now much greater and real-time services have been acquired."

"The redundancy rate assesses the volume of information on a device that has a replication on the Internet (i.e. in computing paradigm servers)."

"In such applications, failure to provide information without internet connectivity in real-time would have a negative effect on system performance."

"This could therefore deliver a real-time Internet connectivity services."

Proposed SD-DRFC Framework Architecture

"Because when the things devices are connected to the DC network, the data initially migrate through dew processing."

"Main purpose and functionalities of DS in this SD-DRFC framework are: Support end-users with services requested, Handle things device connectivity, Control internet connectivity when there is Internet access, Collecting data from things devices, Operated as a digital repeater, as it buffers the emitted stream of data on a low-power radio link and produces a higher-powered signal for transmission to higher-level internet-hosted servers, Storing data to server, Delivery of data to higher level (RC) servers for further storage or analysis, Synchronize and correlate the local and remote data, Things device control, VDC-based data analysis using predefined algorithms, Data visualization, Visualization of the results via User interface."

Simulation Setup and Result Analysis

"Cloud-only and edge-ward (termed "SD-DRFC" in this paper) are 2 placement techniques in this analyzed use case."

"Four QoS parameters are calculated here (Latency, Network Usage, Cost, and Energy Consumption) and the findings are analyzed and presented in terms of the two cases discussed above (i.e. Cloud Only and SD-DRFC)."

"Maybe the most important control loop in this presented use case game is the latency of response which is liable for converting the player's brain state into its game's rules on the smartphone's monitor."

"Because operations (e.g. captured data pre-processing to turn it into information, moderate computing, and data transmission to the upper level) should be performed in the edge level devices in "SD-DRFC" implementation."

"The results show quite a lot of improvement in QoS parameters in the implementation of edge positioning (i.e. "SD-DRFC" implementation) opposed to cloud-only for the IoT framework."

Use Cases

"If the patient is in a location where the internet connection is not available then DC layer still performs the required processing on the data from things device (i.e. fitness tracker) of the patient."

"As the processing performed at the closest possible layer from the data source; hence extreme real-time/real-time services can be provided to patient."

"As the processing task for real-time service is carried out at the edge (i.e. either DC/RC); hence cloud resources are free to perform the time-insensitive big data tasks."

"Upon data gathering from things devices, computing must be done immediately as well as in the nearest possible layer on the time-sensitive data to deliver real-time services."

"To be efficient, the management system (SDN in this case) must perform the computation (i.e. extract processing result when performing on those time-sensitive data) in the closest computation layer and must sending message within fractions of a second to the approaching cars."

Conclusion

"Computing paradigms are such strategies that can be substantially reduced user intervention in IoT ecosystem management."

"In this paper, we have incorporated four computing paradigms (cloud, fog, roof, and dew) to create a multi-tiered IoT ecosystem infrastructure."

"SDN along its network programming abilities, step up as a suitable choice to orchestrate the network, services and devices by covering up the difficulties of this heterogeneous multi-tier computational infrastructure for IoT ecosystem."

"We propose a novel SD-DRFC framework combining multi-tier computational infrastructure (i.e. combining cloud, fog, roof, and dew) which is supported and assisted by SDN."

"The results show quite a lot of improvement in QoS parameters in the implementation of edge positioning (i.e. "SD-DRFC" implementation) opposed to cloud-only for the IoT framework."

"Their improvement, therefore, implies a major advancement in the overall IoT ecosystem QoS. This justifies the novelty and efficiency of our proposed SD-DRFC framework."

Review of the Theory, Principles, and Design Requirements of Human-Centric Internet of Things (IoT) [142]

This is a machine-generated summary of:

Ystgaard, Kaja Fjørtoft; Atzori, Luigi; Palma, David; Heegaard, Poul Einar; Bertheussen, Lene Elisabeth; Jensen, Magnus Rom; De Moor, Katrien: Review of the theory, principles, and design requirements of human-centric Internet of Things (IoT) [142]

Published in: Journal of Ambient Intelligence and Humanized Computing (2023)
Link to original: https://doi.org/10.1007/s12652-023-04539-3

If you want to cite the papers, please refer to the original.

For technical reasons we could not place the page where the original quote is coming from.

Abstract-Summary

"Despite the increased push towards "human-centric" IoT, it is still poorly understood what "human-centric" actually means in this context, how it is interpreted and embedded into the design, by whom, and for which purposes."

"To address these questions, we conducted a systematic literature review (N = 84) on the theory, principles, and design requirements of human-centric IoT. A key observation is that, despite the recent increase in research on humane perspectives for IoT, "human-centredness" often still seems to be used primarily as a label and overarching paradigm, not leading to a profound change in the underlying practices."

"We found no shared understanding of what "human-centric" implies in this context or common agreement on which principles human-centric IoT should be built upon."

"Our findings point towards an apparent discrepancy between how contributions are positioned and framed ("human-centric"), the practices and assumptions they are based on, and their actual impact and ability to orient existing efforts towards genuine human-centric outcomes and key values."

Introduction

"We understand human-centric IoT as an approach that can meet the needs of communities served by IoT technology and that helps to find out what technically works best for humans and society first and foremost."

"IoT-related human-centered design has predominantly had a technical approach, concentrating for example on adding computing to everyday objects, managing connectivity issues for things with dissimilar constraints, technology that allows things being able to communicate to each other, and understanding how data can affect people's sense-making of things [143]."

"To effectively shape a human-centric IoT that establishes human potential in a societal and ethical way, we argue in this paper that the scientific community, engineers, and designers must first develop a shared understanding of what human-centredness conceptually translates to from a socio-technical perspective."

Background

"From the technical point of view, a human-centric IoT design is intended as the process that considers the requirements from the users' point of view in terms of needed functionalities, interaction modalities, and which sensors to adopt to monitor the user status [143]."

"When approached from a socio-technical perspective, human-centered research in IoT should incorporate human involvement via methods used during the system technology design process [144, 145]."

"With the propagation of Cyber-Physical-Systems and IoT into every aspect of humans' everyday life and societal systems, the significance of socio-technical design dimensions in IoT increases [146]: design aspects, such as theoretical approaches, expertise, method for involvement, and technical translation, impact the development of a humanity-centered IoT. We argue in this work that they need to be acknowledged and better understood in order to identify blind spots and to reorient current practices and processes in such a way that key issues can be addressed."

Methodology

"Stage 2: Processing the input; conducting the review, included screening, article selection, review of relevant studies, extraction, and analyses of data."

"Two reviewers screened the article pool for relevance based on the selection criteria."

"A backward and forward search process complemented the article pool (January 2020), searching for articles that either used the selected articles as a reference or were referenced by them [147]."

"From the backward and forward search, we obtained 35 additional articles, and these were screened with the selected criteria, ending up with an addition of 5 articles."

"When duplicates were removed, a total of 88 articles were selected for screening based on pre-existing selection guidelines (inclusion and exclusion criteria), where one author screened first title, keywords and abstract, then two authors screened full-text versions, resulting in an addition of 16 new articles."

Results

"Within the human-centric computing perspective, most of the conceptual definitions of human-centredness focus on describing components or technical resources of IoT using human, social or ethical labelling; examples include Social-internet-of-things, Trusted IoT, Internet-of-people, user-aware IoT or social-aware IoT. A large group of articles defined user-centric as enabling users with tools to protect their data and security."

"The first challenge is the most commonly addressed, with 36 % of the literature looking to guarantee security and privacy (i.e., data protection), describing the intended human-centric design outcomes in terms of security, privacy, and user control."

"We have also considered the study design/contribution type and whether, how, and in what stages users were involved in research under a human-centric IoT umbrella."

Discussion

"A humane and holistic design framework for IoT entails considerations that would work towards safeguarding user rights by regular consultations with diverse audiences in civil society, technical features to flag, prevent or mitigate misuse, and audits where the fault lines of existing power structures are addressed, to ensure equitable and empowering outcomes for all."

"Based on the analyzed papers, the following aspects need to be explicitly considered in future efforts towards building a more human-centric and empowering IoT. In designing for new IoT technologies, the role of non-humans in new IoT/smart environments should be clear and well-understood, especially with regards to defining the relationship with humans."

"There is a mismatch and an overall lack of evaluation of the targeted outcomes to address and manage the effects that IoT systems may have on human users, communities, and society [148, 149]."

Conclusion

"We have sought to address the divide between human-centered/humanity-centred theory and practices in IoT by detailing the existing scope for what human-centered translates into and how it can be practically implemented in the context of IoT. A key observation is that, despite the recent increase in research on humane perspectives for IoT, "human-centredness" often still seems to be used primarily as a label and overarching paradigm, without leading to a profound and necessary change to underlying practices."

"We argue that increased multi-disciplinary research, such as the work presented in this paper, can help to shape how future human-centric IoT roadmaps may realign towards one that benefits humans and humanity first, leading to a successful adoption on a societal scale."

"Future work in this domain must ensure that human-centered IoT is at the core of societal, technological, economic, and political agendas."

Moral Responsibility for Computationally Designed Products [150]

This is a machine-generated summary of:

Douglas, David M.; Howard, David; Lacey, Justine: Moral responsibility for computationally designed products [150]
Published in: AI and Ethics (2021)

Link to original: https://doi.org/10.1007/s43681-020-00034-z

If you want to cite the papers, please refer to the original.
For technical reasons we could not place the page where the original quote is coming from.

Abstract-Summary
"We respond to this problem by distinguishing between causal responsibility and capacity responsibility (the ability to be morally responsible for actions) for creating product designs to argue that while the computational design systems and human designers are both casually responsible for creating product designs, the human designers who use these systems and the developers who create them have capacity responsibility for such designs."

"We show that there is no responsibility gap for products designed using computational design systems by comparing different accounts of moral responsibility for robots and AI (instrumentalism, machine ethics, and hybrid responsibility)."

"We argue that all three of these accounts of moral responsibility for AI systems support the conclusion that the product designers who use computational design systems and the developers of these systems are morally responsible for any flaws or faults in the products designed by these systems."

Introduction

"Designers may see the computational system itself as having full or even partial responsibility for the designed products."

"Clarifying whether the users of a computational design system, the system's developers, or the system itself are responsible for products created using this approach will remove this uncertainty."

"It then describes how the major positions on responsibility for the output of AI systems (instrumentalism, machine ethics, and hybrid responsibility) all converge on attributing moral responsibility to the human users and developers of computational design systems."

"No responsibility gap therefore exists if computational design systems are used to create physical products."

"This paper goes on to explain how the responsibilities of accountability and blameworthiness for design flaws in the products created using computational design may be attributed between the users and developers of these systems, such as computational design systems."

Computational, Parametric, and Generative Design

"Many design packages also come with inbuilt parametric optimisers, which operate on a software representation of the design (a number string with one number per design parameter), and run without human input to maximise one of more fitness functions over a number of iterations, that calculates how well the design performs its intended purpose [151]."

"As these optimisers are only guided at a high level by human-defined performance goals, they have some capacity to provide surprising and non-intuitive solutions (parameter combinations), albeit within a restricted parametric design space."

"Unlike parametric designs, there is a level of indirection between the software representation of the design (the number string), and the complexity of the final design."

"Because of the relationship between number string and final design, and the lack of physically grounded parameters, the design process is much more free-form, with the algorithm searching a wider space than parametric design methods [152, 153, 154]."

Evolutionary Algorithms in Design

"The initialisation stage creates a population where each member is a randomly chosen binary or numerical value or set of values that encodes a possible solution to the design problem."

"If one or more members has a fitness function result that meets the solution's requirements, or if a certain number of iterations of the algorithm have occurred, the genotype with the highest fitness function is converted to its phenotype, which is the problem solution."

"If none of the members produce the required value for the fitness function or if the maximum number of iterations has yet to be reached, the algorithm continues onto the selection stage."

"Description of evolutionary algorithms, we can already identify two points where human design affects the system's output: The selection of representation and associated bounds The definition of the fitness function."

Moral Responsibility and Computational Design

"This loss of influence over the system's decision-making creates uncertainty about whether the developer fulfills the epistemic condition for capacity responsibility for the machine's output."

"Addressing the responsibility gap requires an account of how moral responsibility (i.e., capacity responsibility) should be attributed when the causal responsibility for an AI system's output is uncertain."

"When the system is an implicit ethical agent, machine ethics provides a similar response to that given by instrumentalism: the developers have capacity responsibility for the output of computational design systems."

"Johnson's [155] account of hybrid responsibility describes three relevant entities for a computer system's actions: the human users, the human designers, and the computational system itself."

"For computational design systems, Hanson's and Johnson's accounts agree that the combination of the humans associated with the system and the system itself is causally responsible for the products it designs."

Moral Responsibilities of Human Users and Developers

"The computational design system's human users use it as a tool for achieving the goal of creating new product designs."

"Establishing the specific cause of the fault in the product should indicate whether the fault was due to an error in defining the properties of the search space, a problem in the physical manufacturing process, or due to an unexpected interaction between the materials and the design developed by computational methods."

"Unless the fault is the result of an unforeseen interaction between the materials comprising the product and the designs created via computational design, the human users or developers are wholly accountable for faults in the design of the product and the product itself."

"Human users are blameworthy as product designers if their account reveals that they would be expected to know about the errors and potential risks in the parameters and fitness functions that they defined for the computational design system to use."

Conclusion

"Using computational design to create physical products raises two potential questions for responsibility: does employing computational design create a 'responsibility gap' for product design, and if there is no responsibility gap, who is morally responsible for the products created with the aid of these systems?"

"We have argued that there is no responsibility gap by showing that the human users of computational design systems and the human developers of these systems are morally responsible for the designs these systems produce."

"We hope to demonstrate how the 'responsibility gap' for products designed and made by robots and AI systems may be addressed by considering how responsibility is attributed in specific applications of these technologies."

Software Variability in Service Robotics [156]

This is a machine-generated summary of:

García, Sergio; Strüber, Daniel; Brugali, Davide; Di Fava, Alessandro; Pelliccione, Patrizio; Berger, Thorsten: Software variability in service robotics [156]
Published in: Empirical Software Engineering (2022)
Link to original: https://doi.org/10.1007/s10664-022-10231-5

If you want to cite the papers, please refer to the original.
For technical reasons we could not place the page where the original quote is coming from.

Abstract-Summary
"Engineering robotics software has become increasingly challenging."

"The software needs to be variable—to customize robots for different customers, hardware, and operating environments."

"Variability adds substantial complexity and needs to be managed—yet, ad hoc practices prevail in the robotics domain, challenging effective software reuse, maintenance, and evolution."

"To improve the situation, we need to enhance our empirical understanding of variability in robotics."

"A multiple-case study on software variability in the vibrant and challenging domain of service robotics."

"We investigated drivers, practices, methods, and challenges of variability from industrial companies building service robots."

"We analyzed the state-of-the-practice and the state-of-the-art—the former via an experience report and eleven interviews with two service robotics companies; the latter via a systematic literature review."

"We learned that the level of abstraction in robotics software needs to be raised for simplifying variability management and software integration, while keeping a sufficient level of customization to boost efficiency and effectiveness in their robots' operation."

"To foster software reuse, the service robotics domain will greatly benefit from having software components—completely decoupled from hardware—with harmonized and standardized interfaces, and organized in an ecosystem shared among various companies."

Introduction

"While the drivers and the realization of variability are reasonably well understood in other domains, that is not the case for autonomous robots."

"Our long-term goal is to improve variability management in robotics, where, as we will show, variability is affected by drivers not seen in other domains and is managed in ways lagging behind the state-of-the-art."

"A study on variability management in service robotics."

"We triangulate from three different data sources: our experiences systematically synthesized in an experience report, an in-depth examination of two companies—i.e., a multiple-case study—based on interviews with nine engineers from two robotics companies, and a systematic literature review (SLR)."

"A literature review on variability management in service robotics."

"The most pressing challenge stemming from the identified variability drivers is raising the success rates of configurations and mission specifications that are usable by different robots in various contexts without requiring extensive tuning."

Background

"Robots are cyber-physical systems embodying a blend of hardware and software that interacts with the environment."

"They typically do not require technical knowledge (e.g., programming skills, robotics engineering) from the operator, although they require some training and an expert to program their behaviors beforehand."

"Non-technical operators do not possess programming or robotics engineering knowledge and, therefore, typically resort to end-user-oriented, visual environments to specify missions [157, 158]."

"The implementation of a mission coordinates the robot's skills—programmed actions a robot can perform, often developed as software components by experts—to achieve a mission goal [159–163]."

"Configurable platforms are software systems with variability mechanisms [164–166]."

"Feature models allow keeping an overview understanding of the platform's variability and, together with configurator tools [167–171], allow deriving individual variants in an automated process."

Methodology

"After identifying the main topics we wanted to explore in detail, we designed a multiple-case study and contacted and interviewed nine robotics experts working for two different companies."

"The assessment criteria we used to analyze the search results and record the information that we used to answer our research questions is based on the generated codebook from our multiple-case study."

"The goal was to match our codes from the multiple-case study with the analyzed papers to better triangulate the data between these two sources."

"The comparison revealed a disagreement on applying the inclusion/exclusion criteria to one paper (i.e., whether the paper is relevant) and five disagreements on the interpretation of data to answer our research questions (i.e., what aspects of our research questions were covered by the search results)."

Drivers of Variability (RQ1)

"The operation of service robots needs to consider several scenarios, which compile a set of characteristics of an environment and requirements of the robotic application."

"Standards also affect the definition of the scenario, e.g., by defining the maximum speed a robot can operate in an environment populated by humans."

"Different robot platforms are designed to adapt their behavior based on events from the environment; for instance, any of the studied robots that operate in human-populated environments are expected to avoid collisions with moving objects when navigating."

"Regarding the drivers of variability investigated in our study, environment variability is addressed more explicitly in papers that refer to ground robots as they have to operate in everyday open-ended environments with changing operational conditions (e.g., illumination) that affect the correct acquisition of sensory measurements."

Variability Management Practices (RQ2)

"At PAL Robotics, developers use an in-house variant of SMACH to define finite-state machines (FSMs), which are the mechanism to define the adaptation rules and robotic missions at the company."

"Strategies for managing hardware variability in both studied companies mainly serve two topics, namely developing robotic skills and hardware customization."

"To promote the reusability of skills among platforms as well as hardware customization and software development, a common practice applied by both companies is to harmonize interfaces of software components of their architectures—see the discussion earlier in this section, in the context of robotic skills development."

"A future goal for PAL Robotics is to develop a task planner where high-level missions can be easily defined and which automatically generates a specification based on a target mechanism, namely behavior trees or state machines."

Variability-Related Challenges (RQ3)

"Our interviewees discussed a variety of challenges related to their robots' operating environments, from which we identified (i) conceiving generic solutions for various scenarios, (ii) developing parametric configurations, and (iii) the installation process."

"According to our interviewees, there exist many challenges related to hardware variability in robotics, namely (i) conceiving generic solutions for different robots, (ii) hardware-driven assumptions, (iii) achieving reliability of variant-rich systems, (iv) multiple options for hardware design, (v) lack of standardized hardware options, (vi) performing integration, (vii) aligning conventions for hardware, (viii) trade-offs, (ix) handling real-time components, and (x) influence from environment."

"This simplifies variability management, but, according to one interviewee from PAL Robotics, complicates software development, as explained below."

"One interviewee from PAL Robotics and two from Blue Ocean rate generating the mission that describes the set of goals a robot must achieve while being able to deal with the environment as the most important challenge for robotics variability management."

Discussion

"The characteristics of environment variability mostly discussed in the literature refer to the features and changing conditions of the scenarios where robots operate, similar to some of the findings from our interviews (Obs."

"As opposed, both of our studied companies need to manage the variability stemming from the inclusion of humans to the operating environments of their robots (Obs."

"The variability of robotic software artifacts is driven by the evolution of the underlying technologies in mechanics, electronics, computer science, and cognitive sciences (Obs."

"This means that the variability of robotic software applications is driven by the complexity of everyday environments (Obs."

"This would result in the definition of a common language for expressing the variability in robotic technologies and in robotic requirements that can support robotics engineers in the development of new service robotic applications."

"Robotic companies make use of well-known middleware and frameworks to ease software development (Obs.)"

Threats to Validity

"Minimizing potential threats to the external validity of our study was key for us due to the nature of an SLR, which aims at being as generalizable as possible."

"With this aim, we (i) asked interviewees about practitioner colleagues who may fit in our selection criteria and might be interested in our study, (ii) recruited practitioners with a variety of roles from both companies, and (iii) strove to show our study as theoretically and practically relevant for our potential interviewees."

"To mitigate this bias, we conducted the qualitative analysis of our data collaboratively, that is, the open coding was performed by two authors."

"Another potential threat to conclusion validity is given by the substantial work carried out by a single researcher during the conduction of the SLR."

"The potential bias introduced by this researcher was mitigated by the inclusion of a second researcher during the selection of primary studies and data extraction as quality control."

Related Work

"We confirm some of the variability drivers, variability management practices, and variability challenges already identified in other domains, and we add characteristics that are specific from robotics."

"Berger and others [172] study systematic variability management techniques and software product line engineering (SPLE) concepts in twelve cases that include domains such as automotive, aerospace, or railway systems."

"The authors use a multiple-case study to identify challenges to the adoption of systematic variability management."

"For what concerns challenges, in this work we identified challenges that are specific of the robotic domain, e.g., related to the involvement of non-technical operators who might not be skilled in computer science and/or robotics."

"The latter is an experience report where the authors identify and describe variability issues related to variability management of software product lines."

"This work does not explicitly discuss drivers of variability and challenges."

Conclusion

"We presented an empirical study on variability in service robotics based on the state-of-practice and the state-of-the-art."

"We contribute: (i) characteristics and impacts of drivers of variability in service robotics, (ii) variability management practices applied by service robotics companies, (iii) challenges faced by industrial practitioners as a result of variability in their products, (iv) a discussion of the gap between state-of-practice and state-of-the art, with formulated hypotheses to explain our observations."

"Our results contribute to improving the empirical understanding of the specific variability-related characteristics and challenges of the service robotic domain."

"For robots to operate in several scenarios developers need to both realize variability and integrate robust abstractions."

"Our results give rise to the following future work: Consolidate the drivers of variability in service robotics in a common ontology, acting as a standard."

Internet-Based Intelligent and Sustainable Manufacturing: Developments and Challenges [173]

This is a machine-generated summary of:

Li, Kai; Zhou, Tao; Liu, Bo-hai: Internet-based intelligent and sustainable manufacturing: developments and challenges [173]

Published in: The International Journal of Advanced Manufacturing Technology (2020)
Link to original: https://doi.org/10.1007/s00170-020-05445-0

If you want to cite the papers, please refer to the original.
For technical reasons we could not place the page where the original quote is coming from.

Abstract-Summary
"This paper studies a range of new manufacturing paradigms and presents a state-of-the-art survey of published works."

"It explores the corresponding current manufacturing concepts, technologies, framework features, application effects, resource optimization, and future challenges in these new paradigms."

"Through this survey, the developments of these new manufacturing paradigms are explained and the future prospects are also discussed."

Introduction

"With the increasing personalized demands and market complexity, the rapid updating of emerging information and communication technologies (ICTs) such as cyber-physical system (CPS), Internet of Things (IoT), cloud computing, and big data, and the sustainable development of society, all of these market requirements and technology changes drive the innovation of manufacturing paradigm."

"This paper summarizes the current literature on the advanced manufacturing paradigms and optimization methods of resource utilization, from theoretical and practical developments of emerging information and communication technologies and the Internet platform, into six aspects: (1) intelligent manufacturing, (2) cyber-physical production Systems, (3) industrial Internet of Things, (4) cloud manufacturing, (5) sustainable manufacturing, (6) global manufacturing, (7) mass customization."

"At the end of each section, we summarize and discuss the novel manufacturing paradigm of market demand, product characteristics, customer groups, the requirements to enterprises, the concept of manufacturing model, the features and frameworks of manufacturing systems, the integration of advanced information technologies, the optimization methods of resource utilization, and the application and its future challenges of manufacturing paradigms."

Scope and Methodology

"To clarify advanced manufacturing paradigms and resource optimization methods for manufacturing sector in the internet-based background, this paper presents manufacturing models supported by critical technologies, and discusses the corresponding concepts, market demands, product designs, system development, technology implementation, resource optimization methods, management challenges, application, and development."

"We focus on advanced manufacturing paradigms including intelligent manufacturing, cyber-physical manufacturing systems, industrial Internet of Things, cloud manufacturing, green manufacturing, sustainable manufacturing, global manufacturing, social manufacturing, and mass customization."

"After the initial identification of articles, we conduct a full-text analysis, and ultimately choose the following advanced manufacturing paradigms: intelligent manufacturing, cyber-physical manufacturing systems, industrial Internet of Things, cloud manufacturing, sustainable manufacturing, global manufacturing, and mass customization."

"For each identified manufacturing paradigm, we consider range from the demand analysis, product planning, system development, technology implementation, operation management, resource optimization to application and evolution."

Intelligent Manufacturing

"It makes use of next-generation information, artificial intelligence, manufacturing and management technologies to achieve intelligence of product design, manufacture, management and service [174]."

"Intelligent manufacturing is not only the collection of large amounts of data for the production process, but also data analysis and data mining for guiding the production decision-making and application."

"The combination of multi-agent theory and production scheduling for intelligent manufacturing is one of the approaches to solve the problem of manufacturing process optimization."

"They integrate IoT, big data, artificial intelligence, network technology, and virtual manufacturing technology."

"It is critical for intelligent manufacturing to realize the integration of Internet, network technology, sensing technology, information and communication technology, manufacturing, and management technology."

"Core technologies of Internet-of-Things–based manufacturing are IoT technologies which include sensing technology, network technology, data processing technology, communication technology, artificial intelligence, big data analytics, augmented reality, and virtual reality."

Cyber-Physical Manufacturing

"The implementation of CPS in the manufacturing industry has brought about changes in the manufacturing paradigm and has created cyber-physical production system (CPPS) [175]."

"CPPS consists of autonomous and cooperative elements and subsystems, relying on the latest and foreseeable further developments of computer science, information and communication technology, and manufacturing science and technology, connecting the stages of the production process [176], which are the integration of physical systems and information systems."

"Although CPS in manufacturing is initial stage, with the trends of rapid change of market, the personalized demand of products, reuse of resources, and online service, in the future, CPS in manufacturing will integrate with IoT, cloud computing, and big data, etc. [175]."

"This work proposed a framework for the development of predictive manufacturing cyber-physical systems, based on cloud manufacturing solutions and big data applications."

The Manufacturing Paradigm Driven by the Internet of Things

"In a broad sense, IoT refers to (1) a global network that interconnect intelligent objects by using extended Internet technologies, (2) a set of technologies that support the realization of interconnection, and (3) the new business and market for all applications and services by leveraging such technologies [76, 177]."

"Framework of IoT usually refers to the technology architecture, which includes perception layer for data acquisition at the bottom, the middle of network layer for data transmission, the top application layer for data processing, and the throughout public technology [178]."

"In the adoption of IoT for enterprises, there resides three applications:(1)implementing monitor and control by collecting data, (2) analyzing big data generating from the objects of IoT to provide a reference for business decision-making, and (3) achieving information sharing and collaboration between people and things [177, 179]."

Cloud Manufacturing

"It optimizes the configuration for distributed manufacturing resources, and provides on-demand cloud services."

"Li held that CM was a platform that utilized network and cloud manufacturing services."

"Zhang and others [180] explained that CM was a new manufacturing paradigm that combined emerging information and communication technologies such as cloud computing, IoT, service-oriented technology, and high-performance computing to address bottlenecks in manufacturing applications."

"From the perspective of cloud computing, Xu [181] proposed that CM was cloud computing applied to manufacturing."

"Wang and Xu [182] proposed a service-oriented system called interoperable cloud-based manufacturing system (ICMS) that encapsulated existing and future manufacturing resources by using the virtual function block mechanism and standardized description to provide a cloud-based environment."

"Yu and others [183] regarded CM as an imitation of cloud computing service paradigm, offering on-demand manufacturing services from networks (i.e., cloud-enabled)."

Sustainable Manufacturing Based on Information and Communication Technology

"Deep understanding of sustainable manufacturing and green manufacturing and their mutual links is conducive to improving the use of resources and energy in the production process and reducing the harm to the environment."

"From the microperspective, new manufacturing paradigms have created sustainable manufacturing opportunities for equipment modification, value creation for the organization, organizational change, sustainable design processes, and sustainable product design."

"A process-oriented information model (PIM) was proposed to integrate the information regarding the sustainable manufacturing product design."

"Xu and others [184] proposed a multi-objective joint model of energy consumption and production efficiency for improving the sustainability of manufacturing equipment services in the job shop."

"Nujoom and others [185] developed a multi-objective mathematical model in designing the sustainable manufacturing system for minimizing total economic and environmental costs."

"Manufacturing activities of green manufacturing increase the sustainable use of resources, shorten the product life cycle, and reduce production costs [186]."

Internet-Based Global Manufacturing

"It can complete an integral global manufacturing network by interacting among the organizations."

"This new manufacturing mode would promote socialized resources configuration, social interaction, business collaboration, and all-around production management to accomplish product life cycle efficiently and flexibly."

"One form of value creation in social manufacturing is open production."

"Production master data of the configuration of products, processes, and resources in the global manufacturing network is the basis for design and decision-making of the entire manufacturing network."

"The design and decision-making of global manufacturing networks is becoming more difficult by the dynamics and uncertainties of the business environment."

"Lanza and Moser [187] proposed a dynamic multi-objective optimization model for global manufacturing networks, which evaluated the impact of business environmental factors and optimized production address allocation and capacity planning in the global design of the manufacturing network."

Mass Customization Based on Internet Platform

"MC is also seen as the capability to produce customized products for a mass market [188]."

"The reference models of information systems in existing production and supply chain management cannot meet the requirements of a mass customization driven by dynamic demand."

"Mass customization has been applied to many industries because it can provide a wide variety of products and services to meet the specific requirements of particular customers."

"Silveira and others [189] emphasized the following key success factors for mass customization: the existence of customer demand for variety and customization, the appropriate market conditions, the integral value chain; the available manufacturing and information technology, products should be customizable, knowledge sharing in the value chain."

"MC of automobiles has become an increasing concern for automotive manufacturers, a process chain for the production of shoe for mass customized shoes has been developed, flight catering has adopted most elements of MC, and MC of individualized orthotics also has been developed [190–193]."

Conclusions and Future Prospects

"Enterprises that adopt cloud manufacturing to provide better manufacturing services and gain higher business benefits, need to consider the following: (1) Cloud manufacturing resource cost and profit distribution and business cooperation mode. (2) The cost composition, pricing, bargaining, and operational strategies of cloud service and the corresponding means of electronic payment and security management. (3) The dynamic construction, management, and execution of business process in cloud manufacturing mode. (4) The credit management mechanism for each party (i.e., cloud manufacturing service providers, cloud service demander, cloud manufacturing system brokers). (5) Operational cost optimization of cloud manufacturing service platform."

"Under the sustainable manufacturing environment, manufacturing enterprises need to respond flexibly to the dynamic requirements of customers, to reduce production costs, to consider the ecological environment responsibility of enterprises, to choose to use green raw materials, to form closed-loop product life cycle, to decrease pollutant discharge in the production process, and to create new business models."

Cloud Evolutionary Computation System for Advanced Engineering Analytics [194]

This is a machine-generated summary of:

Chou, Jui-Sheng; Kosasih, Jeffisa Delaosia; Chong, Wai K.: Cloud evolutionary computation system for advanced engineering analytics [194]
Published in: Engineering with Computers (2021)
Link to original: https://doi.org/10.1007/s00366-020-01249-8

If you want to cite the papers, please refer to the original.
For technical reasons we could not place the page where the original quote is coming from.

Abstract-Summary
"The range of applications of artificial intelligence (AI) that is based on nature-inspired metaheuristics is rapidly increasing across various scientific fields as it is used to solve complex engineering problems."

"This work develops a cloud evolutionary machine learning system, called the nature-inspired metaheuristic optimization and prediction system (NiMOPS) that is composed of metaheuristic AI and web modules."

"The objective of the proposed system is to provide a user-friendly web analytics for making efficient, effective, and accurate predictions as solutions to engineering problems."

"A MATLAB Compiler is used to package the system into Java Archive (JAR) files, which provide the core modules for the development of the NiMOPS website using an integrated development environment (IDE)."

"IDE compiles the JAR files, and web utilities (JavaScript, CSS, Servlet, and other utility files) to form the response-request connection between the user and the server."

"The analytic results of using the system to solve classification problems revealed that the system had a fault diagnosis accuracy of 96.5% and an accidental small fire accuracy of 52.4%."

"The numerical experiments verified that cloud computing provides an innovative way to enable decision-makers to solve engineering problems."

Introduction

"The NiMOPS combines a commonly used machine learning (ML) technique with a novel optimization algorithm (OA) to identify model parameters and prevent overfitting and local minima problems."

"LSSVM and LSSVR use a least squares formulation where SVM and SVR formulate and solve a quadratic optimization problem [195]."

"Many researchers are attracted to these algorithms for solving various complex engineering optimization problems [196]."

"Simple and easy to understand, these algorithms are implemented in hybrid applications that can solve very diverse, often highly nonlinear problems."

"This promising algorithm has been tested and found to yield better results than the standard FA and other methods in solving both classification and regression problems [197, 198]."

"The computational costs of solving diverse engineering problems are reduced and a ubiquitously available tool for their solution is provided."

Literature Review

"Chou and Thedja [197] further developed a graphical user interface for the FA with novel metaheuristic agents and integrated the LSSVM to generate a metaheuristic-optimized classification system with enhanced predictive performance."

"This work provides several case studies that validate the use of the developed web-based cloud computing system for solving classification and regression problems in engineering."

"Although many experts and researchers have tried to predict the prices, the predictive performance of housing prices has not attracted much attention because methods of so doing invariably perform poorly [199]."

"To save time and reduce design costs, experts have used AI to model complex nonlinear systems to predict the slump in concrete."

"Various researchers have used ANN techniques and their variations to predict slumps in concrete [196, 200–203], but found that there was still room for improvement on prediction performance."

Methodology and Technique

"MATLAB version R2014a was used as computational software; Java SE Development Kit (JDK) version 1.7.0_79 was used as the development tool and NetBeans version 8.0.2 was used as the integrated development environment (IDE)."

"The Java programming language is the best for use with MATLAB because MATLAB provides a compiler that converts its functions into JAR."

"NetBeans is open source IDE software that is compatible with Java and allows users to code easily, with a range of handy and powerful tools."

"To handle large applications with thousands of folders and files, and millions of lines of code, NetBeans IDE provides various views of the data and multiple project windows to help users set up applications and manage them efficiently."

"When users wish to deploy a web project using NetBeans, the IDE generates a web project main folder and divides it into four folders of web pages, source packages, a library, and configuration files."

Web-Based Computational System Development

"Based on the users' preferences, the system automatically converts the input parameters and the data into text files which also contain outputs."

"In the input form, the "Default" button allows users to use default settings of parameters and the base output file name."

"Users can save the parameter settings and the datasets into a text file in various formats for the four modules using the "Export" button."

"The output allows users, after they have input the parameters and the dataset, to check their input data before performing the analysis."

"When users input the parameter settings and the datasets, the website sends them to Apache Tomcat, which processes all the inputs inside the core system (JAR) until the predicted values have been calculated and the performance measures have been evaluated."

"Users can easily complete the input form, by inputting model settings, a dataset and the output file name."

System Applications

"The dataset of observations comprises 13 input attributes, including one binary variable and the output value, which is the median housing price for the tract."

"The dataset of 103 data was used herein; the seven independent attributes and only one output variable, slump flow (cm), were considered."

"In the prediction of housing prices, the proposed cloud analytics yielded R = 0.92, RMSE = 3.56 (in $1000's), MAE = 2.50 (in $1000's), and MAPE = 13%."

"The prediction performance (RMSE) of the proposed cloud analytics was enhanced by 45.48%, 68.82%, and 37.43% compared with that of bagging, boosting, and clustering methods, respectively."

"Although the performance of the cloud analytics was not better than that of ANN, multiple performance measures revealed that the proposed cloud analytics exhibited more reliably accurate predictions than the other methods."

Conclusions

"This work developed a web-based system for solving engineering problems that can be accessed ubiquitously without the need to purchase computing tools."

"The developed NiMOPS website incorporates the proposed nature-inspired metaheuristic optimization and predictive algorithms with several improvements over main functions that favor simplicity and the user-friendliness of the interface to solve classification and regression problems."

"In estimating home values, the proposed cloud analytics yielded RMSE = 3.56 (in $1000's), which was lower than those of other predictive methods [204, 205]."

"The RMSE of the proposed cloud computing system was 37.43–68.82% better than those of other predictive methods."

"The developed NiMOPS Website enables users to solve engineering problems, providing detailed outputs that are much better than standalone application especially with respect to the graphical results."

"The graphical user interface of web-based cloud computing system can be used without programming knowledge to solve engineering classification and regression problems."

Evolutionary Music: Applying Evolutionary Computation to the Art of Creating Music [206]

This is a machine-generated summary of:

Loughran, Róisín; O'Neill, Michael: Evolutionary music: applying evolutionary computation to the art of creating music [206]
Published in: Genetic Programming and Evolvable Machines (2020)
Link to original: https://doi.org/10.1007/s10710-020-09380-7

If you want to cite the papers, please refer to the original.
For technical reasons we could not place the page where the original quote is coming from.

Abstract-Summary
"A review of the application of genetic programming (GP) and other variations of evolutionary computation (EC) to the creative art of music composition."

"Throughout the development of EC methods, since the early 1990s, a small number of researchers have considered aesthetic problems such as the act of composing music alongside other more traditional problem domains."

"We review the implementation of GP and EC for music composition in terms of the compositional task undertaken, the algorithm used, the representation of the individuals and the fitness measure employed."

"We conclude that the exploratory nature of evolutionary methods are highly appropriate for a wide variety of compositional tasks and propose that the development and study of GP and EC methods on creative tasks such as music composition should be encouraged."

Introduction

"This paper presents a review of the application of GP and more generally EC to music, in particular to the creative act of algorithmic music composition."

"Whereas back in the 1990s practitioners were writing conference papers on such topics, studies and discussions on the application of EC to creative challenges such as music and art have developed into journal articles, chapters and numerous full books focussing on these types of problems [207, 208]."

"The field of Computational Creativity (CC) has emerged and grown in recent years, supporting high-quality research into creative acts including story-telling, art and music."

"We discuss this view, and the suitability of EC to the application of creative acts such as music composition later in this paper."

"GP can be applied to musical or creative systems in any of the above manners."

Evolutionary Music

"In applying EC to music there are a wide range of aspects to consider, however, the three fundamental criteria to consider upfront are problem domain, individual representation and fitness measure."

"A human-based fitness function had been previously used successfully in evolving images [209] but the authors noted the practical difference in that multiple images can be viewed at once—and very quickly—whereas music is a temporal phenomenon and hence all individuals must be listened to sequentially in real time."

"The author noted that while the GA may be simple to implement, it is the fitness measure that is difficult to define and which ultimately determines the musicality of the output."

"An Adaptive Resonant Theory (ART) neural network was used as the fitness measure in a GA applied to musical rhythm composition by Burton and Vladimirova [210]."

Twenty Years of GP Music

"Loughran et al. used the Zipfs distribution among a number of melodic and rhythmic aspects of music in the fitness measure in a system that used GE for the composition of MIDI melodies [211]."

"The fitness measures are developed from a comparison of the notes within the chords and examining the entropy function from partial progressions Ting et al proposed the phrase imitation-based evolutionary composition (PIEC) system which used a GA to create a new melody from the characteristics of a given melodic phrase [212]."

"The authors noted that this would not necessarily create 'better' melodies than their original system, but it did create a complex adaptive system that would generate musical output without relying on an explicit fitness measure from the user."

"MetaCompose was an evolutionary music composer that split the act of composition into three processes: chord generation, melody generation and accompaniment [213]."

Discussion

"Some studies didn't even measure fitness from individuals but instead examined musical trajectories across generations [214] or measured the genetic drift [215] to drive evolution."

"Experiments that use a musical corpus to derive a genre-based fitness measure will always be limited to creating music in a similar style to the given corpus thus limiting the creative potential of the system."

"One other measure that has been used in musical systems but would not be normally seen in EC experiments is the idea of random fitness."

"The interplay of domain, representation and fitness will determine what a given system can and will produce: the search domain determines the space (or 'style') of music, the representation limits the individuals that can be created within that space and the fitness measure determines which individuals will be considered a good result."

"In the context of applying EC methods to music composition it is hence important not to automatically limit evaluations to those taken based on human measurements from the produced musical output."

EC and the Search for Creativity

"We can use evolutionary (and indeed any computational) systems both in generating creative music or in the search for the understanding of creativity."

"It is thus unsurprising that so many systems are emerging that apply, and as we have seen amend, EC algorithms for the purpose of creative studies."

"By providing a way to separate aesthetic judgement from system judgements they proposed that EC methods could offer a better opportunity to explore creative space: The new technologies of GAs and GP offer the promise of tractable evolutionary processing, and hence theories of creativity-through-evolution may now be explored experimentally."

"Searching through this space is not creativity, but searching it effectively could lead to creativity."

"It is this idea of effective search that is imperative for the emergence of creativity from heuristic search methods such as EC."

"The search for creativity relies on effective exploration of the search space."

Conclusion

"We have presented a review on the application of EC methods to the creative task of algorithmic music composition."

"We noted throughout this paper that the application domain of music composition can refer to a large number of musical sub-tasks which must be clearly defined through domain and representation constraints."

"Even when such constraints are in place, one of the biggest challenges for aesthetic applications such as music, in comparison to more traditional problems such as symbolic regression, is that there is no simple way to measure the 'fitness' of a piece of music."

"Music, among the other creative arts, poses a different challenge than numerical tasks in that there is no correct answer."

"Of dwelling on such arguments, we propose to continue the exploration of musical tasks and challenges using the versatility of GP and evolutionary methods in the pursuit of identifying, examining and ideally exhibiting musical creativity."

Membrane Computing Models and Robot Controller Design, Current Results and Challenges [216]

This is a machine-generated summary of:

Buiu, Cătălin; Florea, Andrei George: Membrane computing models and robot controller design, current results and challenges [216]
Published in: Journal of Membrane Computing (2019)
Link to original: https://doi.org/10.1007/s41965-019-00029-8

If you want to cite the papers, please refer to the original.
For technical reasons we could not place the page where the original quote is coming from.

Abstract-Summary
"Designing membrane controllers for single- and multi-robot systems is an application area initiated in 2011 at the Laboratory of Natural Computing and Robotics (natuRO) of the Politehnica University of Bucharest."

"An overview of natuRO's research on the design of robot controllers based on various models of membrane systems is given."

"The main part of this paper is a synopsis of the applications of membrane systems in robot control, while an emphasis is paid to the new membrane computing models introduced at natuRO, Enzymatic Numerical P Systems and XP colonies, and their specific use in single- and multiple-robot applications."

"The paper continues with a critical overview of the performances of membrane controllers as compared to traditional ways to control single- and multiple-robot systems, current challenges and possible ways to overcome these."

Introduction to Robotics and Natural Computing

"In between the sensors and actuators/end-effectors lies the "brain" of the robot, or the robot controller(s)."

"One can say that robotics (in general) is (concerned with the design of) the "intelligent" connection between sensing and action."

"Natural computing paradigms have been proven to be successful in a large number of diverse application areas, among which is the control of robots."

"Fuzzy logic systems have a long tradition of successful applications in designing controllers for robots and autonomous vehicles."

"Neural networks have been successfully applied to the control of robotic manipulators and mobile robots [217–219]."

"Evolutionary algorithms have demonstrated their power and efficiency in optimizing traditional and "intelligent" robot controllers [220–222], including a mobile robot path planning using membrane evolutionary artificial potential field [223]."

"Biomimetics deals with the design of artificial systems (e.g., robots) that are inspired by natural structures that evolved to optimal designs in millions of years of evolution."

Results of Membrane Systems Applied in Robot Control

"D. students affiliated with this laboratory have focused on developing robot control models using MC-based paradigms (ENPS and XPcol were introduced at natuRO): (1) numerical P systems (NPS); (2) enzymatic numerical P systems (ENPS); (3) P colonies (Pcol); (4) XP colonies (XPcol). The first step toward usable MC-based robot controllers was the implementation of a formally correct NPS simulator called SNUPS [224]."

"The control models introduced in [225] successfully handled a homogeneous group of swarm robots, where each robot was controlled by an individual P colony."

"Having obtained an embedded-ready MC robot controller, it was time for a detailed evaluation of these contributions, in direct comparison with simpler models, such as finite state machines and even with model-less applications."

"The results were validated in simulations of up to 50 robots and real swarms of 16 robots and showed that despite the higher model complexity of the proposed XPcol controller, the differences in mean execution time and energy consumption were not significant [226]."

Limitations, Challenges and Solutions

"This was mostly encountered when working with Pcol/XPcol models on Kilobot robots [226–228]."

"The former is necessary because at the very minimum, an entire Pcol program must be loaded in memory for it to be checked whether it is executable."

"The latter is also needed to store the entire set of programs and P objects that make up a Pcol."

"Apart from the memory requirements of the model, there are also those of the Pcol/XPcol simulator that also introduces a considerable overhead [227]."

"Although the XPcol model had a larger execution time than the finite state machine model, this difference was not observable for real robots that have a slow movement speed [226]."

"For applications that require a strict timing, any fluctuations of the execution time of a simulation loop will negatively affect model performance."

Conclusions and Future Work

"The potential of MC in robot control has been explored in a number of challenging applications over the last 10 years, like the multi-robot (or swarm robotics) control."

"Research carried out at natuRO has inspired other groups to follow this direction and explore new PSs variants and to explore new applications in robot control."

"Example avenues for future related works are as follows. (1) Extending the library of NPSs models with new models that are amenable to more complex robot control applications. (2) Implement these models on parallel computing hardware (such as FPGAs) and so extend the application areas to more complicated robots that need faster reaction times while minimizing FPGA memory resource usage and power consumption. (3) An interesting and much needed development for the whole MC community would be the development of an USB device that brings MC inferencing to existing systems, much like an USB accelerator for machine learning applications."

A New Approach to Computing Using Informons and Holons: Towards a Theory of Computing Science [229]

This is a machine-generated summary of:

de la Peña, F. David; Lara, Juan A.; Lizcano, David; Martínez, María Aurora; Pazos, Juan: A New Approach to Computing Using Informons and Holons: Towards a Theory of Computing Science [229]

Published in: Foundations of Science (2019)
Link to original: https://doi.org/10.1007/s10699-019-09597-5

If you want to cite the papers, please refer to the original.
For technical reasons we could not place the page where the original quote is coming from.

Abstract-Summary
"The state of computing science and, particularly, software engineering and knowledge engineering is generally considered immature."

"The best starting point for achieving a mature engineering discipline is a solid scientific theory, and the primary reason behind the immaturity in these fields is precisely that computing science still has no such agreed upon underlying theory."

Introduction

"In the case of computing science (a term that we believe to be better suited than computer science, as it is the science of computing rather than the science of the computer, and hereinafter referred to as CS) and, particularly, software engineering and knowledge engineering, there is a notorious absence of theory (cf. e.g., Hannay and others [230]), to the point that, save very rare exceptions, there are not even any commonly accepted general laws."

"The non-existence of a CS theory contrasts with the everyday reality of other fields of science, where, as indicated in point 2.1, there are indeed established scientific theories on which to found studies, compare experiments or base the respective engineering disciplines."

State of Art and Proposed Theory

"Thinking back to what we said in the last section about the object of study, any theory concerning CS has to consider, that is, establish and define its fundamental element: information and its basic component the "informon"."

"In CS, informons as data interrelate with holons, and informons as news interrelate with the respective holons with which they interact to form what are known as, though inappropriately dubbed, information systems (there is currently a tendency to name some of the existing systems as information systems but a more rigorous definition would be news systems or data systems due to the level of information on which they work)."

"The (holon, informon) pairs and the structures built upon them will inexorably encounter conflicting, concurrent and cooperative situations in which the decision-making schemas they use will be based on both Neumann and Morgenstern's theory [231] and Nash's cooperative game theory [232] and any other decision-making theory or rule that holons can use to make the most appropriate decision at the right time and in the most effective way."

Postulates of the Theory

"This definition takes up the first and explicitly introduces the aspect of information (data) and practical application of the second postulate: the agent (holon) interprets data (information, i.e., informon) and the effects produced in the environment are either a new holon, a new informon, a new holon and a new informon or several holons and/or informons."

"This pair generates the holon Hmtop' as a result, which will be responsible for analyzing the information data base (Ibc') to determine (infer) if it is a fraudulent operation or not by using a framework based system (inferior levels holons residing in the analytic module), where an expert takes part while being supported by LL."

Conclusions

"Unlike other fields of science, where the scientific theories upon which to base the respective engineering disciplines are formally established, CS does not have a formal theory."

"Taking into account the features and peculiarities of this new discipline, this paper has formally established the fundamental elements and postulates that make up a first attempt at a CS theory (a core theory in Lakatos' words [233])."

"The fundamental elements of this theory are informons and holons, and it is formulated as three postulates."

"Bearing this in mind, as well as the fact that the best catalyst for achieving a mature engineering discipline is to have a formal underlying theory, we suggest that both software engineering and knowledge engineering should be founded on the stated theory."

An Innovative Cryptosystem for Needham–Schroeder Authentication Protocols for Tertiary Industries [234]

This is a machine-generated summary of:

Vijayarangan, Avinash; Narayanan, Veena; Srikanth, R.; Vijayarangan, N.: An Innovative Cryptosystem for Needham–Schroeder Authentication Protocols for Tertiary Industries [234]
Published in: Lecture Notes in Mechanical Engineering (2022)
Link to original: https://doi.org/10.1007/978-981-19-0561-2_14

If you want to cite the papers, please refer to the original.
For technical reasons we could not place the page where the original quote is coming from.

Abstract-Summary
"One of the main communication protocols designed for developing robustness across the entire insecure network has been proposed through an improved version of Needham–Schroeder (N-S) protocol."

Introduction

"Applying generalized Pell's equation in Crypto algorithms creates a big challenge in the deduction of private key."

"Sean Hallgren [235] provided polynomial-time quantum algorithms based on Pell's equation."

"Some other studies on Pell's equation in various fields can be seen in [236–241]."

"The Needham–Schroeder Symmetric Key Protocol based on a symmetric encryption algorithm aims to establish a session key between two parties on a network, typically to protect further communication whereas Needham–Schroeder Public-Key Protocol based on public-key cryptography is intended to provide mutual authentication between two parties communicating on a network, but in its proposed form is insecure."

Analysis

"Let A, S, B be the three parties involved in a communication system."

"The flow of communication protocol is shown below."

"In protocol-1, S and A have more or equal priority whereas B has less priority and both members S and A will behave like a blockchain."

"In protocol-2, both A and B will have less priority compared to S who will be a blockchain node."

Applications

"In this situation, A will pass this voucher to the funding institution S which raises the fund for A. As a consequence, S will send a credit note or E-cheque to A who will transfer it to B. In this system, S and B have no mutual communication but A and B have."

"Another banking application is described here for E-cheque transactions from one bank to another."

"Let us have a case study: a payee (A) deposits an E-cheque issued by a drawer (B) through a web application or teller machine."

"Further, both web servers of banks A and B will act as blockchain nodes to invite more payees and drawers for more E-cheques in order to reduce operational complexity and improve the performance."

Conclusion

"The proposed versions of N-S protocols based on the binary quadratic equation have various industrial applications."

"In the field of banking technology, we could design and implement on decentralized ledger for payments, clearance and settlement system, fundraising, tokenizing traditional securities, loans and credit system, trade finance by replacing paper heavy bills of lading process, fraud prevention, mortgage property through financial institutions, and retrieval of borrowers' documents by lenders."

"For all these kinds of transactions, we require a smart execution process and computational robust protocols."

"In due course, we are bringing out implementation results in the banking scenario and improving the existing network performance."

References

1. Aguzzi G, Audrito G, Casadei R, Damiani F, Torta G, Viroli M (2022) A field-based computing approach to sensing-driven clustering in robot swarms. Swarm Intell. https://doi.org/10.1007/s11721-022-00215-y
2. Lin Y, Megerian S (2007) Sensing driven clustering for monitoring and control applications. In: 4th IEEE consumer communications and networking conference, CCNC 2007, Las Vegas, NV, USA, January 11–13, 2007. IEEE, pp 202–206. https://doi.org/10.1109/CCNC.2007.47
3. Viroli M, Beal J, Damiani F et al (2019) From distributed coordination to field calculus and aggregate computing. J Log Algebr Methods Program. https://doi.org/10.1016/j.jlamp.2019.100486
4. Audrito G, Casadei R, Damiani F et al (2021) Optimal resilient distributed data collection in mobile edge environments. Comput Electr Eng. https://doi.org/10.1016/j.compeleceng.2021.107580

5. Alsboui T, Qin Y, Hill R, Al-Aqrabi H (2021) Distributed Intelligence in the Internet of Things: challenges and opportunities. SN Comput Sci. https://doi.org/10.1007/s42979-021-00677-7
6. Alsboui T, Qin Y, Hill R, Al-Aqrabi H (2020) Enabling distributed intelligence in the internet of things with iota and mobile agents. Computing
7. Dizdarevic J, Carpio F, Jukan A, Masip X (2018) A survey of communication protocols for internet of things and related challenges of fog and cloud computing integration. ACM Comput Surv 51(6):1–29
8. Siow E, Tiropanis T, Hall W (2018) Analytics for the internet of things: a survey. ACM Comput Surv 51(4):1–36
9. Noor MM, Hassan WH (2019) Current research on internet of things (IoT) security: a survey. Comput Netw 148:283–294
10. Guan Z, Li J, Wu L, Zhang Y, Du X (2017) Achieving efficient and secure data acquisition for cloud-supported internet of things in smart grid. IEEE Internet Things J 4(6):1934–1944
11. Asif RM, Afsana F, Mahmud M, Shamim KM, Ahmed M, Kaiwartya O, James-Taylor A (2018) Towards a heterogeneous mist, fog, and cloud based framework for the internet of healthcare things. IEEE Internet Things J 10:1
12. Byers CC, Wetterwald P (2015) Fog computing distributing data and intelligence for resiliency and scale necessary for IoT: the internet of things (ubiquity symposium). Ubiquity 4(1–4):12
13. Alsboui T, Qin Y, Hill R (2019) Towards a scalable iota tangle-based distributed intelligence approach for the internet of things. In: Intelligent computing, advances in intelligent systems and computing, vol 10. Springer Verlag
14. Cha S-C, Chen J-F, Su C, Yeh K-H (2018) A blockchain connected gateway for BLE-based devices in the internet of things. IEEE Access 6:24639–24649
15. Fan C, Khazaei H, Chen Y, Musilek P (2019) Towards a scalable dag-based distributed ledger for smart communities. In: 2019 IEEE 5th World Forum on Internet of Things (WF-IoT), pp 177–182
16. Odysseas L, Dionisis P, John G (2019) A novel combination of distributed ledger technologies on internet of things: use case on precision agriculture. Appl Syst Innov 2:30
17. Labrado C, Thapliyal H, Prowell SJ, Teja KP (2019) Use of thermistor temperature sensors for cyber-physical system security. Sensors 19(18):3905
18. Adeleke O, Akinlabi S, Jen TC, Adedeji PA, Dunmade I (2022) Israel Evolutionary-based neuro-fuzzy modelling of combustion enthalpy of municipal solid waste. Neural Comput Applic 34(10):7419–7436. https://doi.org/10.1007/s00521-021-06870-2
19. Qian X, Lee S, Soto AM, Chen G (2018) Regression model to predict the higher heating value of poultry waste from proximate analysis. Resources 7(3):39. https://doi.org/10.3390/resources7030039
20. Olatunji O, Akinlabi S, Madushele N, Adedeji PA (2019) Estimation of Municipal Solid Waste (MSW) combustion enthalpy for energy recovery. EAI Endorsed Trans Energy Web 19(23):1–9. https://doi.org/10.4108/eai.11-6-2019.159119
21. Baghban A, Ebadi T (2019) GA-ANFIS modeling of higher heating value of wastes: application to fuel upgrading. Energy Sources Part A Recover Util Environ Eff 41(1):7–13. https://doi.org/10.1080/15567036.2017.1344746
22. Wang D, Tang YT, He J, Yang F, Robinson D (2021) Generalized models to predict the lower heating value (LHV) of municipal solid waste (MSW). Energy 216:119279. https://doi.org/10.1016/j.energy.2020.119279
23. Xiang F, Fan J, Zhang X, Zuo Y, Liu S (2022) A digital twin-driven perception method of manufacturing service correlation based on frequent itemsets. Int J Adv Manuf Technol. https://doi.org/10.1007/s00170-022-08762-8
24. Lamharhar H, Benhlima L, Chiadmi D (2015) OWLSGOV: an owl-S based framework for e-government services. In: Intelligent distributed computing, vol 321. Springer International Publishing, pp 225–238. https://doi.org/10.1007/978-3-319-11227-5_20

25. Loutas N, Peristeras V, Zeginis D, Tarabanis K (2012) The semantic service search engine (s3e). J Intell Inf Syst 38(3):645–668. https://doi.org/10.1007/s10844-011-0171-6
26. Martinez-González MM, Alvite-Díez ML (2014) On the evaluation of thesaurus tools compatible with the semantic web. J Inf Sci 40(6):711–722. https://doi.org/10.1177/0165551514545603
27. Martino BD (2009) Semantic web services discovery based on structural ontology matching. Int J Web Grid Serv 5(1):46–65. https://doi.org/10.1504/IJWGS.2009.023868
28. Sion R, Tatemura J (2011) Runtime web-service workflow optimization. In: New Frontiers in information and software as services, vol 74. Springer, Berlin, Heidelberg, p 112. https://doi.org/10.1007/978-3-642-19294-4_5
29. Zhang M, Li GX (2017) Multi-objective optimization of manufacturing service composition with interval numbers. Comput Integr Manuf Syst 28(3):1787–1796. https://doi.org/10.13196/j.cims.2017.08.019
30. Tao F, Zhao D, Hu Y, Zhou Z (2010) Correlation-aware resource service composition and optimal-selection in manufacturing grid. Eur J Oper Res 201(1):129–143. https://doi.org/10.1016/j.ejor.2009.02.025
31. Quy VK, Hau NV, Anh DV, Ngoc LA (2021) Smart healthcare IoT applications based on fog computing: architecture, applications and challenges. Complex Intell Syst. https://doi.org/10.1007/s40747-021-00582-9
32. Fu Y, Xiao N, Chen T, Wang J (2021) Fog-to-multicloud cooperative eHealth data management with application-aware secure deduplication. IEEE Trans Depend Secure Comput. https://doi.org/10.1109/TDSC.2021.3086089
33. Sood SK, Mahajan I (2019) IoT-fog-based healthcare framework to identify and control hypertension attack. IEEE Internet Things J 6(2):1920–1927. https://doi.org/10.1109/JIOT.2018.2871630
34. Awaisi KS, Hussain S, Ahmed M, Khan AA, Ahmed G (2020) Leveraging IoT and fog computing in healthcare systems. IEEE Internet Things Mag 3(2):52–56. https://doi.org/10.1109/IOTM.0001.1900096
35. Wang X, Li Y (2020) Fog-assisted content-centric healthcare IoT. IEEE Internet Things Mag 3(3):90–93. https://doi.org/10.1109/IOTM.0001.2000005
36. Isa SBM, El-Gorashi TEH, Musa MOI, Elmirghani JMH (2020) Energy efficient fog-based healthcare monitoring infrastructure. IEEE Access 8:197828–197852. https://doi.org/10.1109/ACCESS.2020.3033555
37. Yang Y, Wang K, Zhang G, Chen X, Luo X, Zhou M (2018) MEETS: maximal energy efficient task scheduling in homogeneous fog networks. IEEE Internet Things J 5(5):4076–4087. https://doi.org/10.1109/JIOT.2018.2846644
38. Arisdakessian S, Wahab OA, Mourad A, Otrok H, Kara N (2020) FoGMatch: an intelligent multi-criteria IoT-fog scheduling approach using game theory. IEEE/ACM Trans Netw 28(4):1779–1789. https://doi.org/10.1109/TNET.2020.2994015
39. Hassija V, Chamola V, Saxena V, Jain D, Goyal P, Sikdar B (2019) A survey on IoT security: application areas, security threats, and solution architectures. IEEE Access 7:82721–82743. https://doi.org/10.1109/ACCESS.2019.2924045
40. Tao H, Bhuiyan MZA, Abdalla AN, Hassan MM, Zain JM, Hayajneh T (2019) Secured data collection with hardware-based ciphers for IoT-based healthcare. IEEE Internet Things J 6(1):410–420. https://doi.org/10.1109/JIOT.2018.2854714
41. Bevish Jinila Y, Merlin Sheeba G, Prayla Shyry S (2021) PPSA: Privacy preserved and secured architecture for internet of vehicles. Wirel Pers Commun. https://doi.org/10.1007/s11277-021-08179-8
42. Nanjie L (2011) Internet of vehicles your next connection. WinWin Magazine 11:23–28. Huawei
43. Golestan K, Soua R, Karray F, Kamel M (2016) Situation awareness within the context of connected cars: a comprehensive review and recent trends. Inform Fusion 29:68–83

44. Tan Z (2012) A light weight conditional privacy preserving authentication and access control scheme for pervasive computing environments. J Netw Comput Appl 35(6):1839–1846
45. Ying B, Makrakis D, Mouftah HT (2013) Privacy preserving broadcast message authentication protocol for VANETs. J Netw Comput Appl 36(5):1352–1364
46. Biswas S, Mišić J (2011) ID-based safety message authentication for security and trust in vehicular networks. In: 31st IEEE International Conference on Distributed Computing Systems Workshops (ICDCSW), pp 323–331
47. Bevish Jinila Y, Komathy K (2014) Cluster oriented ID based multi signature scheme for traffic congestion warning in VANET. In: International conference on emerging ICT for bridging future, vol 338. Springer, pp 337–345
48. HaddadPajouh H, Khayami R, Dehghantanha A, Choo KK, Parizi RM (2020) AI4SAFE-IoT: an AI-powered secure architecture for edge layer of Internet of things. Neural Comput Applic 32(20):16119–16133. https://doi.org/10.1007/s00521-020-04772-3
49. Ren J, Guo H, Xu C, Zhang Y (2017) Serving at the edge: a scalable IoT architecture based on transparent computing. IEEE Netw 31(5):96–105
50. Alaba FA, Othman M, Hashem IAT, Alotaibi F (2017) Internet of Things security: a survey. J Netw Comput Appl 88:10–28
51. Patel P, Ali MI, Sheth A (2017) On using the intelligent edge for IoT analytics. IEEE Intell Syst 32(5):64–69
52. Vučinić M, Tourancheau B, Rousseau F, Duda A, Damon L, Guizzetti R (2015) OSCAR: object security architecture for the Internet of Things. Ad Hoc Netw 32:3–16
53. Pinto S, Gomes T, Pereira J, Cabral J, Tavares A (2017) IIoTEED: an enhanced, trusted execution environment for industrial IoT edge devices. IEEE Internet Comput 21(1):40–47
54. Krithiga R, Geetha P (2020) Deep learning based breast cancer detection and classification using fuzzy merging techniques. Mach Vis Appl. https://doi.org/10.1007/s00138-020-01122-0
55. Sarkar R, Acton ST (2018) SDL: saliency-based dictionary learning framework for image similarity. IEEE Trans Image Process 27(2):749–763
56. Chen H, Qi X, Yu L, Dou Q, Qin J, Heng PA (2017) DCAN: deep contour-aware networks for object instance segmentation from histology images. Med Image Anal 36:135–146
57. Zheng B, Yoon SW, Lam SS (2014) Breast cancer diagnosis based on feature extraction using a hybrid of k-means and support vector machine algorithms. Expert Syst Appl 41(4):1476–1482
58. Oquab M, Bottou L, Laptev I, Sivic J (2014) Learning and transferring mid-level image representations using convolutional neural networks. In: Proceedings of the IEEE conference on computer vision and pattern recognition, pp 1717–1724
59. You YL, Xu W, Tannenbaum A, Kaveh M (1996) Behavioral analysis of anisotropic diffusion in image processing. IEEE Trans Image Process 5(11):1539–1553
60. Wang L, Lu H, Ruan X, Yang MH (2015) Deep networks for saliency detection via local estimation and global search. In: Proceedings of the IEEE conference on computer vision and pattern recognition, pp 3183–3192
61. Park KT, Lee J, Kim HJ, Noh S (2020) Digital twin-based cyber physical production system architectural framework for personalized production. Int J Adv Manuf Technol 106(5):1787–1810. https://doi.org/10.1007/s00170-019-04653-7
62. Monostori L et al (2016) Cyber-physical systems in manufacturing. CIRP Ann Manuf Technol 65:621–641
63. Ribeiro L, Björkman M (2017) Transitioning from standard automation solutions to cyber-physical production systems: an assessment of critical conceptual and technical challenges. IEEE Syst J:1–13
64. Ribeiro L (2017) Cyber-physical production systems' design challenges. In: IEEE ISIE. IEEE, pp 1189–1194
65. Brecher C et al (2012) Integrative production technology for high-wage countries. In: Integrative production technology for high-wage countries. Springer, Berlin, pp 17–76

66. Ding K et al (2019) Defining a digital twin-based cyber-physical production system for autonomous manufacturing in smart shop floors. Int J Prod Res:1–20
67. Tao F et al (2018) Digital twin-driven product design, manufacturing and service with big data. Int J Adv Manuf Technol 94:3563–3576
68. Vachálek J et al (2017) The digital twin of an industrial production line within the industry 4.0 concept. In: IEEE PC. IEEE, pp 258–262
69. Zhang H et al (2017) A digital twin-based approach for designing and multi-objective optimization of hollow glass production line. IEEE Access 5:26901–26911
70. Farahani S, et al. (2019). Numerical simulation for the hybrid process of sheet metal forming and injection molding using smoothed particle hydrodynamics method. No. 2019-01-0713. SAE Technical Paper, 2019
71. Cheng Y et al (2018) Cyber-physical integration for moving digital factories forward towards smart manufacturing: a survey. Int J Adv Manuf Technol:1–13
72. Alam KM, El Saddik A (2017) C2ps: a digital twin architecture reference model for the cloud-based cyber-physical systems. IEEE Access 5:2050–2062
73. Rojas RA, Rauch E (2019) From a literature review to a conceptual framework of enablers for smart manufacturing control. Int J Adv Manuf Technol:1–17
74. Gabor T et al (2016) A simulation-based architecture for smart cyber-physical systems. IEEE ICAC:374–379
75. Park KT et al (2019) Service-oriented platform for smart operation of dyeing and finishing industry. Int J Comput Integr Manuf 32(3):307–326
76. Lee J, Bagheri B, Kao H-A (2015) A cyber-physical systems architecture for industry 4.0-based manufacturing systems. Manuf Lett 3:18–23
77. Liu Q et al (2018) Digital twin-driven rapid individualised designing of automated flow-shop manufacturing system. Int J Prod Res:1–17
78. Zhuang C, Liu J, Xiong H (2018) Digital twin-based smart production management and control framework for the complex product assembly shop-floor. Int J Adv Manuf Technol 96(1–4):1149–1163
79. Tao F, Zhang M (2017) Digital twin shop-floor: a new shop-floor paradigm towards smart manufacturing. IEEE Access 5:20418–20427
80. Da Xu L, He W, Li S (2014) Internet of things in industries: a survey. IEEE Trans Ind Informat 10:2233–2243
81. Papazoglou MP et al (2007) Service-oriented computing: state of the art and research challenges. Computer 40
82. Xia L, Lu J, Zhang H, Xu M, Li Z (2022) Construction and application of smart factory digital twin system based on DTME. Int J Adv Manuf Technol 120(5):4159–4178. https://doi.org/10.1007/s00170-022-08971-1
83. Parri J, Sampietro S, Vicario E (2018) Deploying digital twins in a lambda architecture for industry 4.0. ERCIM NEWS 115:30–31
84. Parri J, Patara F, Sampietro S, Vicario E (2020) A framework for model-driven engineering of resilient software-controlled systems. Computing. https://doi.org/10.1007/s00607-020-00841-6
85. Gabor T, Belzner L, Kiermeier M, Beck MT, Neitz A (2016) A simulation-based architecture for smart cyber-physical systems. In: 2016 IEEE International Conference on Autonomic Computing (ICAC). IEEE, pp 374–379
86. Schmidt DC, Stal M, Rohnert H, Buschmann F (2013) Pattern-oriented software architecture, patterns for concurrent and networked objects, vol 2. Wiley, Hoboken
87. Ericson CA (1999) Fault tree analysis. In: System safety conference, Orlando, Florida, vol 1, pp 1–9
88. Schmidt DC (2006) Model-driven engineering. Comput IEEE Comput Soc 39(2):25
89. Akilandeswari J, Jothi G, Naveenkumar A, Sabeenian RS, Iyyanar P, Paramasivam ME (2022) Design and development of an indoor navigation system using denoising autoencoder based

convolutional neural network for visually impaired people. Multimed Tools Appl. https://doi.org/10.1007/s11042-021-11287-z
90. Lin S, Cheng R, Wang K, Yang K (2018) Visual localizer: outdoor localization based on convnet descriptor and global optimization for visually impaired pedestrians. Sensors 18(8):2476
91. Mehmood F, Ullah I, Ahmad S et al (2019) Object detection mechanism based on deep learning algorithm using embedded IoT devices for smart home appliances control in CoT. J Ambient Intell Hum Comput. https://doi.org/10.1007/s12652-019-01272-8
92. Poncela A, Coslado F, García B et al (2018) Smart care home system: a platform for eAssistance. J Ambient Intell Hum Comput. https://doi.org/10.1007/s12652-018-0979-9
93. Kirchner N, Furukawa T (2005) Infrared localisation for indoor uavs. In: Proc 1st Intern Conf Sens Technol
94. Yayan U, Yucel H (2015) A low cost ultrasonic based positioning system for the indoor navigation of mobile robots. J Intell Rob Syst 78(3–4):541–552
95. Medina C, Segura JC, De la Torre A (2013) Ultrasound indoor positioning system based on a low-power wireless sensor network providing sub-centimeter accuracy. Sensors 3:3501–3526
96. Khowaja SA, Khuwaja P (2021) Q-learning and LSTM based deep active learning strategy for malware defense in industrial IoT applications. Multimed Tools Appl 80(10):14637–14663. https://doi.org/10.1007/s11042-020-10371-0
97. Arshad J, Azad MA, Abdeltaif MM, Salah K (2020) An intrusion detection framework for energy constrained IoT devices. Mech Syst Signal Process 136:106436. https://doi.org/10.1016/j.ymssp.2019.106436
98. Laskey M, Staszak S, Hsieh WY-S et al (2016) SHIV: reducing supervisor burden in dagger using support vectors for efficient learning from demonstrations in high dimensional state spaces. In: 2016 IEEE International Conference on Robotics and Automation (ICRA). IEEE, pp 462–469
99. Zhang J, Cho K (2016) Query-efficient imitation learning for end-to-end autonomous driving
100. Moskovitch R, Nissim N, Englert R, Elovici Y (2008) Active learning to improve the detection of unknown computer worms activity. In: 11th International conference on information fusion, pp 1–8
101. Liu Y (2004) Active learning with support vector machine applied to gene expression data for Cancer classification. J Chem Inf Comput Sci 44:1936–1941. https://doi.org/10.1021/ci049810a
102. Pang K, Dong M, Wu Y, Hospedales T (2018) Meta-learning transferable active learning policies by deep reinforcement learning
103. Hassen M, Carvalho MM, Chan PK (2017) Malware classification using static analysis based features. In: IEEE Symposium Series on Computational Intelligence (SSCI). IEEE, pp 1–7
104. Afonso VM, de Amorim MF, Grégio ARA, Junquera GB, de Geus PL (2015) Identifying android malware using dynamically obtained features. J Comput Virol Hacking Tech 11:9–17. https://doi.org/10.1007/s11416-014-0226-7
105. Sharmeen S, Huda S, Abawajy JH, Ismail WN, Hassan MM (2018) Malware threats and detection for industrial mobile-IoT networks. IEEE Access 6:15941–15957. https://doi.org/10.1109/ACCESS.2018.2815660
106. Naeem H, Ullah F, Naeem MR, Khalid S, Vasan D, Jabbar S, Saeed S (2020) Malware detection in industrial internet of things based on hybrid image visualization and deep learning model. Ad Hoc Netw 105:102154. https://doi.org/10.1016/j.adhoc.2020.102154
107. Wang Z, Schaul T, Hessel M et al (2016) Dueling network architectures for deep reinforcement learning. In: Proceedings of the 33rd international conference on machine learning, pp 1995–2003
108. Mnih V, Kavukcuoglu K, Silver D et al (2013) Playing atari with deep reinforcement learning. In: Proceedings of the Conference on Neural Information Processing Systems (NIPS), workshop on deep learning, pp 1–9
109. Mnih V, Badia AP, Mirza M et al (2016) Asynchronous methods for deep reinforcement learning. In: Proceedings of the 33rd international conference on machine learning, pp 1928–1937

110. Dipsis N, Stathis K (2019) A RESTful middleware for AI controlled sensors, actuators and smart devices. J Ambient Intell Human Comput. https://doi.org/10.1007/s12652-019-01439-3
111. Kakas A, Mancarella P, Sadri F, Stathis K, Toni F (2008) Computational logic foundations of KGP agents. J Artif Intell Res 33(1):285–348
112. Witkowski M, Stathis K (2004) A dialectic architecture for computational autonomy. In: Nickles M, Rovatsos M, Weiss G (eds) Agents and computational autonomy. autonomy, Lecture notes in Computer Science, vol 2969. Springer, Berlin
113. Bellifemine FL, Caire G, Greenwood D (2007) developing multi-agent systems with Jade. Wiley, Hoboken
114. ZigBee (2019) ZigBee wireless standard. https://www.zigbee.org/. Accessed 10 Jul 2019
115. Arduino (2017). http://www.arduino.cc. Accessed 30 Sep 2017
116. Google Assistant (2017). https://assistant.google.com. Accessed 30 Sep 2017
117. Amazon Alexa (2017). http://alexa.amazon.com/. Accessed 30 Sep 2017
118. IFTTT (2019) If this then that. https://ifttt.com/. Accessed 7 May 2019
119. Pradhan MR, Mago B, Ateeq K (2022) A classification-based sensor data processing method for the internet of things assimilated wearable sensor technology. Clust Comput. https://doi.org/10.1007/s10586-022-03605-3
120. Manogaran G, Alazab M, Saravanan V, Rawal BS, Sundarasekar R, Nagarajan SM, Kadry S, Montenegro-Marin CE (2020) Machine learning assisted information management scheme in service concentrated IoT. IEEE Trans Industr Inform 17:2871–2879
121. Manogaran G, Shakeel PM, Fouad H, Nam Y, Baskar S, Chilamkurti N, Sundarasekar R (2019) Wearable IoT smart-log patch: an edge computing-based Bayesian deep learning network system for multi access physical monitoring system. Sensors 19(13):3030
122. Guo X, Lin H, Wu Y, Peng M (2020) A new data clustering strategy for enhancing mutual privacy in healthcare IoT systems. Future Gener Comput Syst 113:407–417
123. Billah MFRM, Saoda N, Gao J, Campbell B (2021) BLE can see: a reinforcement learning approach for RF-based indoor occupancy detection. In: Proceedings of the 20th international conference on information processing in sensor networks (co-located with CPS-IoT Week 2021), pp 132–147
124. Liao H, Zhou Z, Zhao X, Zhang L, Mumtaz S, Jolfaei A, Ahmed SH, Bashir AK (2019) Learning-based context-aware resource allocation for edge-computing-empowered industrial IoT. IEEE Internet Things J 7(5):4260–4277
125. Mukherjee R, Ghorai SK, Gupta B, Chakravarty T (2020) Development of a wearable remote cardiac health monitoring with alerting system. Instrum Exp Tech 63:273–283
126. Sarmah SS (2020) An efficient IoT-based patient monitoring and heart disease prediction system using deep learning modified neural network. IEEE Access 8:135784–135797
127. Fouad H, Mahmoud NM, El Issawi MS, Al-Feel H (2020) Distributed and scalable computing framework for improving request processing of wearable IoT assisted medical sensors on pervasive computing system. Comput Commun 151:257–265
128. Zhou H, Montenegro-Marin CE, Hsu CH (2020) Wearable IoT based cloud assisted framework for swimming persons in health monitoring system. Curr Psychol. https://doi.org/10.1007/s12144-020-00822-0
129. Wu T, Wu F, Qiu C, Redoute JM, Yuce MR (2020) A rigid-flex wearable health monitoring sensor patch for IoT-connected healthcare applications. IEEE Internet Things J 7:6932–6945
130. Yacchirema D, Sarabia-Jácome D, Palau CE, Esteve M (2018) System for monitoring and supporting the treatment of sleep apnea using IoT and big data. Pervasive Mob Comput 50:25–40
131. Rammuni Silva RS, Fernando P (2022) Effective utilization of multiple convolutional neural networks for chest X-ray classification. SN Comput Sci 3(6):492. https://doi.org/10.1007/s42979-022-01390-9
132. Porpodis K et al (2014) Pneumothorax and asthma. J Thorac Dis 6:1. Available: http://jtd.amegroups.com/article/view/2173

133. Rosenkrantz AB, Hughes DR, Duszak R (2016) The US radiologist workforce: an analysis of temporal and geographic variation by using large national datasets. Radiology 279(1):175–184. https://doi.org/10.1148/radiol.2015150921
134. Silva RRS, Fernando P (2021) An extensive survey of machine learning based approaches on automated pathology detection in chest X-rays'. In: 2021 28th Conference of Open Innovations Association (FRUCT), vol 9. IEEE, pp 365–373
135. Dai, W., et al.: SCAN: structure correcting adversarial network for organ segmentation in chest X-rays, ArXiv170308770 Cs, 2017, Accessed: 02 Aug 2020. Available: http://arxiv.org/abs/1703.08770
136. Liu H, Wang L, Nan Y, Jin F, Wang Q, Pu J (2019) SDFN: segmentation-based deep fusion network for thoracic disease classification in chest X-ray images. Comput Med Imaging Graph 75:66–73. https://doi.org/10.1016/j.compmedimag.2019.05.005
137. Ravuri HK, Vega MT, van der Hooft J, Wauters T, De Turck F (2021) A scalable hierarchically distributed architecture for next-generation applications. J Netw Syst Manage. https://doi.org/10.1007/s10922-021-09618-4
138. Bannour F, Souihi S, Mellouk A (2018) Distributed SDN control: survey, taxonomy, and challenges. IEEE Commun Surv Tutor 20(1):333–354
139. Ravuri HK, Torres Vega M, van der Hooft J, Wauters T, Da B, De Turck F (2020) On routing scalability in flat SDN architectures. In: NoF2020, the 11th international conference on network of the future, pp 1–5
140. Bonfim MS, Dias KL, Fernandes SF (2019) Integrated NFV/SDN architectures: a systematic literature review. ACM Comput Surv 51(6):1–39
141. Ahammad I, Khan MA, Salehin ZU (2021) Software-defined dew, roof, fog and cloud (SD-DRFC) framework for IoT ecosystem: the journey, novel framework architecture, simulation, and use cases. SN Comput Sci. https://doi.org/10.1007/s42979-021-00521-y
142. Ystgaard KF, Atzori L, Palma D, Heegaard PE, Bertheussen LE, Jensen MR, De Moor K (2023) Review of the theory, principles, and design requirements of human-centric Internet of Things (IoT). J Ambient Intell Human Comput 14(3):2827–2859. https://doi.org/10.1007/s12652-023-04539-3
143. Koreshoff TL, Leong TW, Robertson T (2013a) Approaching a human-centered Internet of Things. In: Proceedings of the 25th Australian computer-human interaction conference: augmentation, application, innovation, collaboration, pp 363–366
144. Breve B, Desolda G, Deufemia V, Greco F, Matera M (2021) An end-user development approach to secure smart environments. In: International symposium on end user development. Springer, Cham, pp 36–52
145. Ghiani G, Manca M, Patern (2017) Personalization of context-dependent applications through trigger-action rules. ACM Trans Comput Hum Interact 24(2):1–33
146. Streitz N, Charitos D, Kaptein M, Böhlen M (2019) Grand challenges for ambient intelligence and implications for design contexts and smart societies. J Ambient Intell Smart Environ 11(1):87–107
147. Levy Y, Ellis TJ (2006) A systems approach to conduct an effective literature review in support of information systems research. Inform Sci 9:181–212
148. Memedi M, Tshering G, Fogelberg M, Jusufi I, Kolkowska E, Klein G (2018) An interface for IoT: feeding back health-related data to Parkinson's disease patients. J Sens Actuator Netw 7(1):14
149. Shin DH, Park YJ (2017) Understanding the Internet of Things ecosystem: multi-level analysis of users, society, and ecology. Digit Policy Regul Gov 19(1):77–100
150. Douglas DM, Howard D, Lacey J (2021) Moral responsibility for computationally designed products. AI Ethics 1(3):273–281. https://doi.org/10.1007/s43681-020-00034-z
151. Kicinger R, Arciszewski T, de Jong K (2005) Evolutionary computation and structural design: a survey of the state-of-the-art. Comput Struct 83:1943–1978. https://doi.org/10.1016/j.compstruc.2005.03.002

152. Law ELC, Van Schaik P (2010) Modelling user experience-an agenda for research and practice. Interact Comput 22(5):313–322
153. Corbin JM, Strauss A (1990) Grounded theory research: procedures, canons, and evaluative criteria. Qual Soc 13(1):3–21
154. Corbin J, Strauss A (2014) Basics of qualitative research: techniques and procedures for developing grounded theory. Sage Publications
155. Johnson DG (2006) Computer systems: moral entities but not moral agents. Ethics Inf Technol 8(4):195–204. https://doi.org/10.1007/s10676-006-9111-5
156. García S, Strüber D, Brugali D, Di Fava A, Pelliccione P, Berger T (2022) Software variability in service robotics. Empir Softw Eng. https://doi.org/10.1007/s10664-022-10231-5
157. Dragule S, Berger T, Menghi C, Pelliccione P (2021a) A survey on the design space of end-user oriented languages for specifying robotic missions. Int J Softw Syst Model (SoSYM)
158. Ajaykumar G, Steele M, Huang CM (2021) A survey on end-user robot programming. ACM Comput Surv 54(8)
159. García S, Pelliccione P, Menghi C, Berger T, Bures T (2019a) High-level mission specification for multiple robots. In: Proceedings of the 12th ACM SIGPLAN international conference on software language engineering, pp 127–140
160. Ghzouli R, Berger T, Johnsen EB, Dragule S, Wasowski A (2020) Behavior trees in action: a study of robotics applications. In: 13th ACM SIGPLAN International Conference on Software Language Engineering (SLE)
161. Menghi C, Tsigkanos C, Berger T, Pelliccione P, Ghezzi C (2018) Property specification patterns for robotic missions. In: Proceedings of the 40th international conference on software engineering: companion proceedings, pp 434–435
162. Menghi C, Tsigkanos C, Pelliccione P, Ghezzi C, Berger T (2019) Specification patterns for robotic missions. IEEE Trans Softw Eng
163. Dragule S, Garcia S, Berger T, Pelliccione P (2021b) Languages for specifying missions of robotic applications. In: Cavalcanti A, ad Rob Hierons BD, Timmis J, Woodcock J (eds) Software engineering for robotics. Springer
164. Apel S, Batory D, Kästner C, Saake G (2013a) Feature-ppes. Springer, Berlin
165. Van der Linden FJ, Schmid K, Rommes E (2007) Software product lines in action: the best industrial practice in product line engineering. Springer Science & Business Media
166. Berger T, Pfeiffer RH, Tartler R, Dienst S, Czarnecki K, Wasowski A, She S (2014) Variability mechanisms in software ecosystems. Inf Softw Technol 56(11):1520–1535
167. Bashroush R, Garba M, Rabiser R, Groher I, Botterweck G (2017) Case tool support for variability management in software product lines. ACM Comput Surv 50(1)
168. Krueger CW (2007) Biglever software gears and the 3-tiered SPL methodology. In: Companion to the 22nd ACM SIGPLAN conference on object-oriented programming systems and applications companion, pp 844–845
169. Kastner C, Thum T, Saake G, Feigenspan J, Leich T, Wielgorz F, Apel S (2009) Featureide: a tool framework for feature-oriented software development. In: Proceedings of the 31st international conference on software engineering, ICSE'09
170. Hubaux A, Jannach D, Drescher C, Murta L, Männistö T, Czarnecki K, Heymans P, Nguyen T, Zanker M (2012) Unifying software, product configuration: a research roadmap. In: Proceedings of the workshop on configuration at ECAI 2012; Montpellier, France, August 27, 2012. CEUR-WS, vol 958, pp 31–35
171. Franz P, Berger T, Fayaz I, Nadi S, Groshev E (2021) Configfix: interactive configuration conflict resolution for the linux kernel. In: 43rd International Conference on Software Engineering, Software Engineering in Practice Track (ICSE/SEIP)
172. Berger T, Steghöfer JP, Ziadi T, Robin J, Martinez J (2020) The state of adoption and the challenges of systematic variability management in industry. Empir Softw Eng 25:1755–1797
173. Li K, Zhou T, Liu BH (2020) Internet-based intelligent and sustainable manufacturing: developments and challenges. Int J Adv Manuf Technol. https://doi.org/10.1007/s00170-020-05445-0

174. Oztemel E, Gursev S (2020) Literature review of industry 4.0 and related technologies. J Intell Manuf 31(1):127–182
175. Wang L, Törngren M, Onori M (2015) Current status and advancement of cyber-physical systems in manufacturing. J Manuf Syst 37:517–527
176. Monostori L, Kádár B, Bauernhansl T, Kondoh S, Kumara S, Reinhart G, Sauer O, Schuh G, Sihn W, Ueda K (2016) Cyber-physical systems in manufacturing. CIRP Ann-Manuf Techn 65(2):621–641
177. Miorandi D, Sicari S, De Pellegrini F, Chlamtac I (2012) Internet of things: vision, applications and research challenges. Ad hoc Netw 10(7):1497–1516
178. Beal J, Pianini D, Viroli M (2015) Aggregate programming for the internet of things. Computer 48(9):22–30. https://doi.org/10.1109/MC.2015.261
179. Lee I, Lee K (2015) The Internet of Things (IoT): applications, investments, and challenges for enterprises. Bus Horiz 58(4):431–440
180. Zhang L, Luo Y, Tao F, Li BH, Ren L, Zhang X, Guo H, Cheng Y, Hu A, Liu Y (2014) Cloud manufacturing: a new manufacturing paradigm. Enterp Inform Syst 8(2):167–187
181. Xu X (2012) From cloud computing to cloud manufacturing. Robot Comp-Int Manuf 28(1):75–86
182. Wang XV, Xu XW (2013) ICMs: a cloud-based manufacturing system. In: Cloud manufacturing. Springer, pp 1–22
183. Yu C, Xu X, Lu Y (2015) Computer-integrated manufacturing, cyber-physical systems and cloud manufacturing–concepts and relationships. Manuf Lett 6:5–9
184. Xu W, Shao L, Yao B, Zhou Z, Pham DT (2016) Perception data-driven optimization of manufacturing equipment service scheduling in sustainable manufacturing. J Manuf Syst 41:86–101
185. Nujoom R, Wang Q, Mohammed A (2018) Optimisation of a sustainable manufacturing system design using the multi-objective approach. Int J Adv Manuf Technol 96(5–8):2539–2558
186. Paul ID, Bhole GP, Chaudhari JR (2014) A review on green manufacturing: it's important, methodology and its application. Procedia Mater Sci 6:1644–1649
187. Lanza G, Moser R (2014) Multi-objective optimization of global manufacturing networks taking into account multi-dimensional uncertainty. CIRP Ann-Manuf Techn 63(1):397–400
188. Wang Z, Zhang M, Sun H, Zhu G (2016) Effects of standardization and innovation on mass customization: an empirical investigation. Technovation 48:79–86
189. Silveira GD, Borenstein D, Fogliatto FS (2001) Mass customization: literature review and research directions. Int J Prod Econ 72(1):1–13
190. Alford D, Sackett P, Nelder G (2000) Mass customisation—an automotive perspective. Int J Prod Econ 65(1):99–110
191. Chang YYC, Jones P (2007) Flight catering: an investigation of the adoption of mass customisation. J Hosp Tour Manag 14(1):47–56
192. Denkena B, Scherger S (2005) A concept for shoe last manufacturing in mass customisation. CIRP Ann-Manuf Techn 54(1):341–344
193. Kaiser C, Fischer TV, Schmeltzpfenning T, Stöhr M, Artschwager A (2014) Case study: mass customisation of individualized orthotics–the fashion-able virtual development and production framework. Procedia CIRP 21:105–110
194. Chou JS, Kosasih JD, Chong WK (2021) Cloud evolutionary computation system for advanced engineering analytics. Eng Comput. https://doi.org/10.1007/s00366-020-01249-8
195. Kisi O, Parmar KS (2016) Application of least square support vector machine and multivariate adaptive regression spline models in long term prediction of river water pollution. J Hydrol 534:104–112. https://doi.org/10.1016/j.jhydrol.2015.12.014
196. Sahab MG, Toropov VV, Gandomi AH (2013) A review on traditional and modern structural optimization: problems and techniques. In: Hossein A, Xin-She G, Siamak Y, Amir T, Alavi H (eds) Metaheuristic applications in structures and infrastructures. Elsevier, Oxford, pp 25–47. https://doi.org/10.1016/B978-0-12-398364-0.00002-4

197. Chou J-S, Thedja JPP (2016) Metaheuristic optimization within machine learning-based classification system for early warnings related to geotechnical problems. Autom Constr 68:65–80. https://doi.org/10.1016/j.autcon.2016.03.015
198. Chou J-S, Yang K-H, Pampang JP, Pham A-D (2015) Evolutionary metaheuristic intelligence to simulate tensile loads in reinforcement for geosynthetic-reinforced soil structures. Comput Geotech 66:1–15
199. Knight JR, Hill RC, Sirmans C (1992) Biased prediction of housing values. Real Estate Econ 20(3):427–456
200. Yeh IC (1998) Modeling of strength of high-performance concrete using artificial neural networks. Cem Concr Res 28(12):1797–1808. https://doi.org/10.1016/S0008-8846(98)00165-3
201. Yeh I-C (2006) Exploring concrete slump model using artificial neural networks. J Comput Civ Eng 20(3):217–221. https://doi.org/10.1061/(ASCE)0887-3801(2006)20:3(217)
202. Yeh IC (2007) Modeling slump flow of concrete using second-order regressions and artificial neural networks. Cement Concr Compos 29(6):474–480. https://doi.org/10.1016/j.cemconcomp.2007.02.001
203. Sobhani J, Najimi M, Pourkhorshidi AR, Parhizkar T (2010) Prediction of the compressive strength of no-slump concrete: a comparative study of regression, neural network and ANFIS models. Construct Build Mater 24(5):709–718
204. Williams CK (1997) Regression with Gaussian processes. In: Mathematics of neural networks. Springer Science + Business Media, New York, pp 378–382. https://doi.org/10.1007/978-1-4615-6099-9
205. Qiang F, Shang-Xu H, Sheng-Ying Z (2005) Clustering-based selective neural network ensemble. J Zhejiang Univ Sci A 6(5):387–392. https://doi.org/10.1007/BF02839405
206. Loughran R, O'Neill M (2020) Evolutionary music: applying evolutionary computation to the art of creating music. Genetic Program Evol Mach. https://doi.org/10.1007/s10710-020-09380-7
207. Miranda ER, Al Biles J (2007) Evolutionary computer music. Springer, Berlin
208. Romero JJ (2008) The art of artificial evolution: a handbook on evolutionary art and music. Springer, Berlin
209. Sims K (1993) Interactive evolution of equations for procedural models. Vis Comput 9(8):466–476
210. Burton AR, Vladimirova T (1998) Genetic algorithm utilising neural network fitness evaluation for musical composition. In: Smith GD et al (eds) Artificial neural nets and genetic algorithms. Springer, Berlin, pp 219–223
211. Loughran R, McDermott J, O'Neill M (2015) Grammatical evolution with zipf's law based fitness for melodic composition. In: Sound and music computing conference, Maynooth
212. Ting CK, Wu CL, Liu CH (2015) A novel automatic composition system using evolutionary algorithm and phrase imitation. IEEE Syst J 11:1284
213. Scirea M, Togelius J, Eklund P, Risi S (2016) Metacompose: a compositional evolutionary music composer. In: International conference on evolutionary and biologically inspired music and art. Springer, pp 202–217
214. Eigenfeldt A, Pasquier P (2012) Populations of populations: composing with multiple evolutionary algorithms. In: Evolutionary and biologically inspired music, sound, art and design. Springer, Berlin, pp 72–83
215. de Freitas AR, Guimaraes FG, Barbosa RV (2012) Ideas in automatic evaluation methods for melodies in algorithmic composition. In: Sound and music computing conference
216. Buiu C, Florea AG (2019) Membrane computing models and robot controller design, current results and challenges. J Membr Comput 1(4):262–269. https://doi.org/10.1007/s41965-019-00029-8
217. Fang W, Chao F, Yang L, Lin CM, Shang C, Zhou C et al (2019) A recurrent emotional CMAC neural network controller for vision-based mobile robots. Neurocomputing 334:227–238. https://doi.org/10.1016/J.NEUCOM.2019.01.032

218. Luan F, Na J, Huang Y, Gao G (2019) Adaptive neural network control for robotic manipulators with guaranteed finite-time convergence. Neurocomputing 337:153–164. https://doi.org/10.1016/J.NEUCOM.2019.01.063
219. Williams HA, Jones MH, Nejati M, Seabright MJ, Bell J, Penhall ND et al (2019) Robotic kiwifruit harvesting using machine vision, convolutional neural networks, and robotic arms. Biosyst Eng 181:140–156. https://doi.org/10.1016/J.BIOSYSTEMSENG.2019.03.007
220. Assis L, Soares ADS, Coelho CJ, Van Baalen J (2016) An evolutionary algorithm for autonomous robot navigation. Procedia Comput Sci 80:2261–2265. https://doi.org/10.1016/J.PROCS.2016.05.404
221. Haiek DE, Aboulissane B, Bakkali LE, Bahaoui JE (2019) Optimal trajectory planning for spherical robot using evolutionary algorithms. Procedia Manuf 32:960–968. https://doi.org/10.1016/J.PROMFG.2019.02.309
222. Larsen L, Schuster A, Kim J, Kupke M (2018) Path planning of cooperating industrial robots using evolutionary algorithms. Procedia Manuf 17:286–293. https://doi.org/10.1016/J.PROMFG.2018.10.048
223. Orozco-Rosas U, Montiel O, Sepúlveda R (2019) Mobile robot path planning using membrane evolutionary artificial potential field. Appl Soft Comput 77:236–251. https://doi.org/10.1016/J.ASOC.2019.01.036
224. Arsene O, Buiu C, Popescu N (2011) SNUPS—a simulator for numerical membrane computing. Int J Innov Comput 7(6):3509–3522
225. Florea AG, Buiu C (2016) Development of a software simulator for P colonies. Applications in robotics. Int J Unconvent Comput 12(2–3):189–205
226. Florea AG, Buiu C (2018) A distributed approach to the control of multi-robot systems using XP colonies. Integr Comput-Aid Eng 25(1):15–29. https://doi.org/10.3233/ICA-170554
227. Florea AG (2018) Contributions to the control of collective robotic systems using membrane computing. PhD thesis. Politehnica University of Bucharest, Bucharest, Romania
228. Florea AG, Buiu C (2017) Membrane computing for distributed control of robotic swarms: emerging research and opportunities. IGI Global, USA. https://doi.org/10.4018/978-1-5225-2280-5
229. de la Peña FD, Lara JA, Lizcano D, Martínez MA, Pazos J (2019) A new approach to computing using informons and holons: towards a theory of computing science. Found Sci. https://doi.org/10.1007/s10699-019-09597-5
230. Hannay JE, Sjøberg DIK, Dybå T (2007) A systematic review of theory use in software engineering experiments. IEEE Trans Softw Eng 33(2):87–107
231. von Neumann J, Morgenstern O (1953) Theory of games and economic behavior. Princeton University Press, New Jersey
232. Nash JF (1953) Two person cooperative game. Econometrica 21:128–140
233. Lakatos I, Musgrave A (1970) Criticism and the growth of knowledge. Cambridge University Press, Cambridge
234. Vijayarangan A, Narayanan V, Srikanth R, Vijayarangan N (2022) An innovative cryptosystem for Needham–Schroeder authentication protocols for tertiary industries. In: Industry 4.0 and advanced manufacturing. https://doi.org/10.1007/978-981-19-0561-2_14
235. Hallgren S (2002) Polynomial-time quantum algorithms for Pell's equation and the principal ideal problem. In: Proceedings of the thirty-fourth annual ACM symposium on theory of computing, pp 653–658
236. Mishra AM (2014) A digital signature scheme based on Pell equation. Int J Innov Res Sci Eng Technol 3(1)
237. Chakravarthy ASN, Balaji S, Koduri R (2012) Evaluation of watermarking techniques. Int J Math Trends Technol 3
238. McBride A (1999) Remarks on Pell's equation and square root algorithms. Math Gaz 83(496):47–52
239. Raghunandan KR, Shetty R, Aithal G (2017) Key generation and security analysis of text cryptography using cubic power of Pell's equation. In: 2017 International Conference

on Intelligent Computing, Instrumentation and Control Technologies (ICICICT). IEEE, pp 1496–1500
240. Raghunandan KR, Dsouza RR, Rakshith N, Shetty S, Aithal G (2021) Analysis of an enhanced dual RSA algorithm using Pell's equation to hide public key exponent and a fake modulus to avoid factorization attack. In: Advances in artificial intelligence and data engineering. Springer, Singapore, pp 809–823
241. Segar TC, Vijayaragavan R (2013) Pell's RSA key generation and its security analysis. In: 2013 Fourth International Conference on Computing, Communications and Networking Technologies (ICCCNT). IEEE, pp 1–5

Chapter 3
Industrial Challenges and Solutions

Introduction by the Editor

In the modern era of industry, data modeling, computations and methods play a major role in the commercial world. There is a challenge in implementing computations and methods for bringing out products and solutions to customers in the cutting-edge industry of health care, aerospace, retail, hotels, financial institutions and manufacturing due to demographic and behavioral issues. After assimilating CAMS and computation methods, readers could start thinking towards open problems in industry. This chapter presents challenges and solutions for medical image denoising, EEG signals for digit classification, water quality monitoring, image encryption, fault diagnosis of rotating machinery, cardiac abnormalities and simulation models. Industries have started solving challenges using generalized and greedy methods through CAMS, which serves data models using a cluster of artificial intelligence algorithms, prediction systems for customers' preferences and computing dynamic offers.

In this chapter, we bring that real-time medical image denoising and information hiding using deep wavelet multiscale autonomous unmanned analysis is a complex and cutting-edge area of research at the intersection of medical imaging, signal processing, and artificial intelligence. Latest research on electroencephalogram (EEG) signals has been using CAMS methods. Low-energy real Fast Fourier Transform (FFT) architectures of CAMS play a crucial role in processing electroencephalogram (EEG) signals for seizure prediction due to their ability to efficiently analyze frequency components of EEG data while minimizing power consumption. Then quantitative tissue characterization by means of discrete wavelet transform (DWT) of photoacoustic signals exploits the potential of this method or system for analyzing biological tissues.

In this chapter, we are discussing a research article on a new image encryption algorithm using discrete orthogonal Stockwell transform (DOST) and fractional

V. Natarajan (ed.), *Computational Artificial Intelligence and Methods for industries*, https://doi.org/10.1007/978-981-96-5277-8_3

Fourier transform (FrFT). It is an interesting topic for readers. The symmetric encryption or private key algorithms that leverage transforms like DOST and FrFT can offer unique advantages in terms of security and computational efficiency. Next, we exploit methods to monitor the dynamic response of bridges using wavelet transform techniques, particularly focusing on cross-correlation wavelet analysis. The innovative encryption method seems quite advanced and effective for analyzing signals from different types of excitations (traffic-dependent and impulsive energizers) applied to bridge structures. In general, wavelet transforms indeed offer valuable tools for denoising signals and analyzing their frequency content with time localization.

Next, we propose the wavelet analysis to mechanical devices such as dynamic response of bridges. Through investigating or analyzing the mechanical response of a bridge using an innovative impulsive energizer combined with wavelet analysis is a fascinating approach that can provide valuable insights into the structural dynamics and integrity of the bridge.

In this chapter, we have focused on solving acoustic emission (AE). Examine the condition of grinding wheels through AE signals is crucial for ensuring machining quality and preventing costly failures. That is the reason, an efficient short-time Fourier transform (STFT) algorithm can be a potential instrumental in analyzing these AE signals to assess the health of the grinding wheel. Further. We discuss a research paper that compares different FFT architectures for real-valued FFT computations, specifically for low-throughput applications like biomedical signals. The research study uses Xilinx Ultra96-V2 FPGA development board to implement and compare single processing element (SPE), pipelined, and in-place architectures. As a consequence, the research findings highlight the trade-offs in terms of throughput, resource utilization, and energy consumption, especially showing that pipelined architectures offer higher throughput but may consume more power compared to in-place architectures for certain applications like seizure prediction systems.

Next, we dive into the monitoring of the grinding process and the characteristics of grinding wheels using acoustic emission (AE) signals. It is proven that AE technology indeed offers valuable insights into the condition and performance of grinding operations, including monitoring cutting edges during grinding and dressing operations. Besides, we explore and exploit research algorithms to monitor the surface integrity of grinding wheels using acoustic emission (AE) signals, particularly through time-frequency analysis using STFT. This method allows for better understanding of AE signal characteristics during operations like dressing, providing insights into the behavior of frequencies over time.

In this chapter, we bring out the challenges related to electronic components. Classifying motherboard components using electromagnetic (EM) side-channel signals through deep learning involves a multi-step process that combines data collection, data preprocessing, signal processing, feature extraction, model training, deep learning model selection and validation.

In this chapter, we have highlighted the challenges on solving bearing faults. Let us introduce a novel intelligent approach based on Weighted Overlapping Adaptive Generalized Weighted Variational Mode Decomposition (WOAGWO-VMD) and

Multi-Perspective Analysis Least Squares Support Vector Machine (MPA-LSSVM) for diagnosis of bearing faults is a sophisticated method that operates or combines advanced signal processing and machine learning techniques to accurately monitor, detect and classify faults in bearings. Then the research work promotes an advanced fault-diagnosis approach for rolling bearings to mitigate catastrophic failures and economic losses. The inventive method leverages a combination or mixture of novel optimization algorithms and machine learning techniques or methods to handle the complexity of vibration signals, which are often nonlinear and nonstationary. Further, the chapter describes Hybrid Optimization Algorithms (HOA) like WOAGWO-VMD which facilitates 3 main features: Dispersion Entropy (DE), Permutation Entropy (PE) and Singular Value Decomposition (SVD).

Let us discuss the applications of deep learning in machinery. The recent advances in the machinery fault diagnosis, the deep learning is applied to the rotating machinery using vibration signals that have shown significant progress in enhancing or improving the accuracy, efficiency, and reliability of fault detection and condition monitoring systems. Due to the latest developments, fault diagnosis in bevel gearboxes using Coiflet wavelet and fault classification based on Artificial Neural Networks (ANN) including Deep Neural Networks (DNN) is a comprehensive approach aimed at accurately detecting and classifying faults in the gearbox as discussed in this chapter.

In this chapter, let us present the challenging task or open issues like the task of detecting epileptic seizures by classifying electroencephalography (EEG) signals into ictal (seizure) and interictal (seizure-free) states. Then the non-linear and non-stationary nature of EEG signals as discussed above complicates the extraction of comprehensive information necessary for accurate classification. How to avoid the complication during EEG signals processing? For EGG analysis, K-Nearest Neighbor (KNN) and Fuzzy Rough Nearest Neighbor (FRNN) are applied to synthesize the signals and extract them in a better way. The integration of fuzzy-based and traditional machine learning or deep learning algorithms in the research framework enhances the detection accuracy of epileptic seizures by effectively handling the complex characteristics of EEG signals. Further, the applications of Generative Artificial intelligence in healthcare (GenAI) have come into the picture. Is there a way to avoid cardiac abnormalities due to a sudden heart attack? Nowadays, electrocardiogram (ECG) signals for cardiac abnormalities have been identified by GenAI algorithms for the analysis of electrocardiogram (ECG). Hence CAMS offers an improved accuracy, efficiency, reliability and scalability for healthcare products or systems with the advancement of GenAI algorithms. Today's scenario, the healthcare service providers encourage "the use of CAMS based ECG systems" to control prevailing heart attacks. Let us discuss cardiovascular diseases (CVDs) that are the prime or leading cause of mortality both in India and globally, as per the report provided by the World Health Organization (WHO). Due to ischemic heart disease, anemia, hypertension, smoking, hypersexuality, alcohol intake, and stressful lifestyles, there are irregularities in the heart's rhythm of our human body, and it will lead to cardiac arrhythmias (one type of CVDs). Thus, CAMS will resolve CVDs. Additionally, long-term mental stress and anxiety percolating or affecting the Autonomic Nervous System (ANS) can also lead to cardiac arrhythmias, posing or threatening a significant challenge for researchers and the healthcare industry.

To expand implementation of simulation-based decision-support systems to the execution phase of project planning stages, CAMS solves effectively using simulation models. The use of Bayesian inference with Markov chain Monte Carlo (MCMC)–based numerical approximation approach solves the project planning at different scenarios and guides the decision-support systems.

Machine Generated Summaries

Disclaimer: The summaries in this chapter were generated from Springer Nature publications using extractive AI auto-summarization: An extraction-based summarizer aims to identify the most important sentences of a text using an algorithm and uses those original sentences to create the auto-summary (unlike generative AI). As the constituted sentences are machine selected, they may not fully reflect the body of the work, so we strongly advise that the original content is read and cited. The auto generated summaries were curated by the editor to meet Springer Nature publication standards. To cite this content, please refer to the original papers.

Machine generated keywords: signal, EEG, image, wavelet, vibration, transform, fault, diagnosis, frequency, electrical, spectral, noise, soft, imaging, wavelet transform.

Real-Time Medical Image Denoising and Information Hiding Model Based on Deep Wavelet Multiscale Autonomous Unmanned Analysis [1]

This is a machine-generated summary of:

Guo, Guangxing: Real-time medical image denoising and information hiding model based on deep wavelet multiscale autonomous unmanned analysis [1]
Published in: Soft Computing (2022)
Link to original: https://doi.org/10.1007/s00500-022-07322-2

If you want to cite the papers, please refer to the original.
For technical reasons we could not place the page where the original quote is coming from.

Abstract-Summary
"Medical image provides a more intuitive visual means, which provides a reliable basis for disease diagnosis, clinical treatment and teaching research."

"In real life, many different kinds of noise will affect the image collection process, these factors will make the quality of the image to a large extent."

"Of image denoising, white noise is a very common kind of noise."

"Even a high-resolution image will have some noise, but some simple denoising algorithms can be used for these noises."

"We study the real-time medical image denoising and information hiding model based on deep wavelet multiscale autonomous unmanned analysis."

"The experimental results show that this method can effectively deal with the noise points in the medical image, and the result graph is better."

Introduction

"When we denoise the medical image, we need to eliminate the influence of noise and save the effective boundary information of the image."

"In the frequency domain method, the information is mainly concentrated in the middle and high frequency parts of the noise after the Gaussian distribution is transformed in the frequency domain, then the appropriate frequency domain filter is selected for filtering processing, and then the denoised image is obtained after the inverse transformation."

"The main problems of medical image denoising are as follows: (1) it is only effective for Gaussian noise, but not for impulse noise."

"These methods have improved the medical image processing technology in different degrees, and the details and edge information of the image have been better protected."

"This paper constructs a medical image denoising model based on wavelet multiscale analysis algorithm."

Wavelet Multiscale Analysis

"In less than 20 years since the concept of wavelet was put forward, wavelet transform has made a breakthrough in theory and method, and has been widely applied in many fields such as signal analysis and image processing, data compression, computer vision, speech synthesis and analysis."

"The wavelet analysis can analyze the signal in time and frequency domain at the same time, and has the function of auto zoom."

"The basic idea of wavelet image threshold denoising: when the wavelet coefficient is less than a certain threshold, the wavelet coefficient is caused by noise and is eliminated."

"When the wavelet coefficient is larger than a certain threshold value, it is caused by the actual signal to retain it. (1) A group of wavelet coefficients are obtained by selecting wavelet bases, decomposing layers and decomposing noisy images. (2) According to the properties of wavelet transform, image wavelet coefficients are composed of useful coefficients and noise coefficients."

Construction of Medical Image Denoising System

"The protection of image edge contour is higher than linear filtering at the same scale."

"In the general sense, median filtering is relative to linear filtering to protect the edge and contour information of the image."

"The median filter removes the noise of the image, and the details of the image must be lost."

"When the image contains more details, the standard median filtering algorithm is not suitable."

"The algorithm can automatically select the appropriate window size, the small window can protect the image details completely, and the large window can effectively filter out the noise."

"Compared with the ordinary median filter, the adaptive median filter can process the image with higher noise density, and can save the edge and details of the image completely when filtering the image polluted by non-impulse noise."

Experiment and Analysis

"SNLM keeps more edge details than NLM in the region with more edge changes."

"This experiment verifies the similarity between the original image and its rotating image."

"The original undistorted image is slowly rotated by 3 degrees."

"Of rotation, a value is taken every 0.3 degrees as a distorted image."

Conclusion

"Medical image denoising, medical image segmentation and contour extraction are the basis of medical image processing."

"In medical image processing, denoising is very important."

"It is the basis of image recognition, image segmentation, contour extraction and other subsequent operations."

"Wavelet analysis is widely used in many fields of image processing, including image compression, image enhancement, image denoising and so on."

"A medical image denoising model based on wavelet multiscale analysis is constructed."

"The experimental results show that the method proposed in this paper has good recognition effect and can accurately distinguish the feature points in medical images."

EEG Signals to Digit Classification Using Deep Learning-Based One-Dimensional Convolutional Neural Network [2]

This is a machine-generated summary of:

Tiwari, Smita; Goel, Shivani; Bhardwaj, Arpit: EEG Signals to Digit Classification Using Deep Learning-Based One-Dimensional Convolutional Neural Network [2]
Published in: Arabian Journal for Science and Engineering (2022)
Link to original: https://doi.org/10.1007/s13369-022-07313-3

If you want to cite the papers, please refer to the original.
For technical reasons we could not place the page where the original quote is coming from.

Abstract-Summary
"The EEG signals were isolated into an energy–frequency–time spectrum with Hilbert Huang transform that was used by the Deep Learning (DL)-based model to learn discriminative spectro-temporal patterns of the raw EEG signals of ten digits."

"This paper has two major contributions: first, create a novel dataset known as BrainDigiData of EEG signals of ten digits from (0–9) using a multi-channel EEG device."

"To propose a DL-based one-dimensional Convolutional neural network model BrainDigiCNN to classify the BrainDigiData of EEG signals of digits."

"The publicly available Mind Big Dataset (MBD) of digits was also used to evaluate the performance of the proposed model."

"The research done in this paper showed that the band-wise analysis of EEG signals in a complex scenario resulted in improved results as compared to the scenario used in the previously existing work for digit classification using EEG signals."

Introduction

"EEG-based motor-imagery tasks were classified by analyzing the signals using EMD [3], and EEG signals for multiple classes were analyzed, and features were extracted using EMD for Brain-Computer Interface (BCI) [4, 5] The data collected from the EEG device are time-series data."

"Such applications using the DL-based CNN model for EEG signal classification motivated the research done in this paper."

"The 1D CNN model performed well in the analysis of the time-varying EEG signal data."

"The novelty and major contributions of this paper are as follows: A novel DL-based model known as the "BrainDigiCNN" model was proposed for the feature extraction and classification of EEG signals of digits from (0–9)."

"A novel DL-based model "BrainDigiCNN" model was implemented to extract the discriminative spectro-temporal patterns of the raw EEG signals in the form of IA, IP and IF of the ten digits and classify the EEG signals of digits from (0–9) using the extracted multi-channel features."

Related Work

"The ML-based KNN classifier was used to classify the EEG signals of digits from (0–9) [6]."

"The publicly available MBD was used to classify EEG signals of digits from (0–9)."

"The Fast Fourier Transform (FFT) was combined with the PCA method to extract features from the EEG signals of the digits from 0 to 9 using the MBD dataset."

"The average accuracy obtained using a weighted KNN classifier for classifying EEG signals of digits was 84%."

"The publicly available EEG data of digits were collected using a five-channel EEG device EMOTIV Insight for one subject."

"Three different DL models based on CNN, Gated Recurrent Unit (GRU), and autoencoder were implemented to extract features and then classify the EEG signals of digits from (0–9) [7]."

"The author created an EEG-based dataset of the digits from (0–9) using a 14-channel EEG device at a sampling rate of 128 Hz."

Materials and Methods

"The prior instruction for the data collection was given to all subjects before the recording of their EEG signals of the brain corresponding to each of the stimulus digits."

"Data preparation One of the most important tasks of this research was to create and collect a local dataset of EEG signals corresponding to the stimulus of digit (0–9)."

"It includes a description of the signal processing technique, feature extraction using EMD and HHT, and the DL-based 1D CNN model to classify the EEG signals of digits and the adopted hyperparameters for it."

"The EMD was applied to the transposed raw EEG signals of the digits dataset in each frequency band."

"The extracted features were passed as an input to the 1D CNN classifier for the classification of EEG signals of digits from (0–9)."

"The signal processing was done on a public MBD dataset of EEG signals of digits from (0–9) for one subject."

Experimental Results

"The proposed model achieved an average classification accuracy of 97.2%, 97.7%, 96.36%, and 96.83% for EEG signals of theta, alpha, beta high, and gamma bands, respectively."

"The proposed BrainDigiCNN model compared to the traditional ML algorithms achieved the highest average accuracy of 96.99% and 98.08% across all six sub-bands for BrainDigiData and MBD datasets, respectively."

"The BrainDigiCNN model reported the lowest accuracy of 95.03% for classifying EEG signals of the beta low band for BrainDigiData and the lowest accuracy of 97.24% for classifying EEG signals of the beta high band of the MBD dataset."

"The performance of the proposed BrainDigiCNN model on the MBD dataset is found to be better as compared to previous work done for classifying EEG signals of digits."

"The classification results on the MBD dataset of ten digits achieved the highest results of EEG signal classification using the proposed BrainDigiCNN model as compared to the existing works."

Conclusion

"A novel approach for digit classification using EEG signals is proposed in this research paper."

"The main purpose of this research was to create new EEG data of digits from (0–9) and propose a model for the classification of EEG signals in a complex scenario having more classes along with big data of EEG signals."

"A novel model BrainDigiCNN model for EEG-based digit classification is proposed."

Calcium Soft Sensor Based on the Combination of Support Vector Regression and 1-D Digital Filter for Water Quality Monitoring [8]

This is a machine-generated summary of:

Dilmi, Smail: Calcium Soft Sensor Based on the Combination of Support Vector Regression and 1-D Digital Filter for Water Quality Monitoring [8]
Published in: Arabian Journal for Science and Engineering (2022)
Link to original: https://doi.org/10.1007/s13369-022-07263-w

If you want to cite the papers, please refer to the original.
For technical reasons we could not place the page where the original quote is coming from.

Abstract-Summary
"Soft sensors are an efficient and economical way to solve such problems."

"Conventional soft sensors are subject to several external factors that may affect the robustness of the sensor, the most important of which is the noise generated by the measurement process."

"A calcium (Ca) soft sensor is developed with lower costs and more robustness with respect to noise."

"Support Vector Regression (SVR), Extreme Learning Machine (ELM), Elman Recurrent Neural Network (ERNN), and Multilayer perceptron (MLP) models are used to develop the soft sensor."

"Principal Component Analysis (PCA) technique is used to choose the variables that have the greatest impact on the soft sensor."

"The results showed that the developed models outperform the traditional soft sensors."

Introduction

"The development of intelligent systems for water quality monitoring remains a major challenge due to the complexity of a large number of parameters and the difficulty related to their measurement [9, 10]."

"The complicated behavior of parameters measured by each sensor presents a major challenge in water quality monitoring [9, 11]."

"This water quality parameter sensor is very high cost and completely unreliable."

"Haghiabi and others [12] presented a comparative study of three ML techniques, including SVM, group method of data handling (GMDH), and artificial neural network (ANN) for predicting water quality parameters."

"ELM technique was used to develop a Cl soft sensor for water quality monitoring [13]."

"Qishlaqi and others [14] developed an MLP model to predict water quality parameters of the Tireh River (Iran)."

"Their study indicated that the MLP model has a suitable accuracy to predict the water quality parameters."

Proposed Water Quality Monitoring System

"Soft sensors are used to solve many problems, including the high price, the use of a large number of sensors, the cleaning and calibration processes, and the long time dedicated to inspecting and replacing the sensors."

"The objective of using an LPF is to eliminate high-frequency noise."

"Ensuring a good performance of the soft sensor with high robustness concerning the noise."

"Through this approach, the failure output of the sensor and/or some of the laboratory sensors can be reproduced."

Methods

"There are two primary families of digital filters: the first, the finite impulse response (FIR) filters; and the second, the infinite impulse response (IIR) filters."

"The FIR filter is distinguished from the IIR filter by its ease of design and high speed."

"The IIR filter is very popular in real-time applications such as biomedical sensor signal processing and IoT smart sensors [15, 16]."

"An IIR filter is a type of electronic filter characterized by a response based on the values of the input signal as well as the previous values of that same response."

"The generalization performance of an SLFN is improved by reducing training errors as well as the norm output weights."

"The connections between neurons are called weights, at first, their values are assigned randomly, and then they are improved during the training phase."

Evaluation Metrics

"The lower the RMSE, the better the performance of the model; the MAE metric is a commonly understood parameter in the applications of engineering."

"The lower the MAE, the better the performance of the model; the MSE metric evaluates the mean squared difference between the measured and expected values."

"The lower the MSE, the better the performance of the model; the RRSE metric indicates how well the model performs relative to the mean actual values."

"The lower the RRSE, the better the model; the R^2 metric represents the degree of correlation between the measured and expected values."

Results and Discussion

"We note that the IIR-ERNN model gives the best performance when using three hidden layers with 5 neurons for each layer (RMSE = 2.493 mg/L, MAE = 1.955 mg/L, MSE = 6.216 mg/L, RRSE = 31.65%, and $R^2 = 0.899$), while the IIR-MLP model gives the best performance when using one hidden layer with 5 neurons (RMSE = 2.540 mg/L, MAE = 2.020 mg/L, MSE = 6.453 mg/L, RRSE = 32.59%, and $R^2 = 0.893$)."

"By comparing the performance of the models in the two cases (i.e., with and without filtering the input data), in general, the performance of the standalone SVR model was better compared to the other standalone models (i.e., ELM, ERNN, and MLP), this is on the one hand."

"A Ca soft sensor was developed based on the integration of four ML techniques (i.e., ELM, ERNN, SVR, and MLP) with an IIR digital filter."

Conclusion and Future Work

"This paper presents a contribution to the development of a soft sensor with lower costs and more robustness concerning noise by integrating ML techniques with an IIR digital LPF for measuring Ca concentration in the field of water quality monitoring."

"SVR, ELM, ERNN, and MLP techniques were used to develop the soft sensor."

"The PCA technique was used to choose the variables which have the greatest impact on the soft sensor."

"The results showed that the developed models (i.e., IIR-SVR, IIR-ELM, IIR-ERNN, and IIR-MLP) outperform the traditional soft sensors, i.e., the standalone models (SVR, ELM, ERNN, and MLP)."

Image Encryption Using Discrete Orthogonal Stockwell Transform with Fractional Fourier Transform [17]

This is a machine-generated summary of:

Ranjan, Rajeev; Thakur, Abhishek: Image encryption using discrete orthogonal Stockwell transform with fractional Fourier transform [17]
Published in: Multimedia Tools and Applications (2022)
Link to original: https://doi.org/10.1007/s11042-022-14240-w

If you want to cite the papers, please refer to the original.
For technical reasons we could not place the page where the original quote is coming from.

Abstract-Summary
"A new image encryption algorithm has been proposed for image security."

"It has been used for image encryption and decryption."

"Encryption of the image has been used DOST with independent specific keys and its fractional-order and analyzed the effect of DOST."

"Decrypted the image through the inverse process of the image encryption."

"The encrypted image has been achieved using the correct order and keys."

"Calculate the correlation coefficient of the original, encryption image, encryption time, and entropy."

Introduction

"It is necessary to explore the image encryption technique."

"This technique is a way to transmit and receive the personal data of the image."

"Several image encryption techniques have been studied; one proposed method is based on DOST with FrFT."

"Illustrate the comparative analysis of DOST and DWT techniques for recognizing the medical images."

"Four image encryption scheme is described by combining the use of the quaternion Fresnel transform (QFST), Computer-generated hologram (CGH), and 2-D logistic adjusted sine map techniques [18]."

"The image encryption with essential techniques is helpful in the real-time encryption of image applications."

"Stenography is a technique in which images, messages, or file data are hidden inside the image, message, or file."

"Visual cryptography (VC) is a technique used to encrypt visual data such as images."

"A 3D chaotic map to design an image encryption technique and chaotic-based partial image technique is discussed in [19, 20]."

S-Transform

"The signal x(t) is defined in the ST domain like X(τ, f), and it is expressed as [21]."

"The Gaussian window g(t - τ, f)is controlled by time shift 'τ' and frequency 'f'."

"τ', it gives the FT of the signal x(t), and is expressed as The condition of normalization of a Gaussian window is defined as [22]."

"The discrete FT (DFT) is expressed as [23, 24]."

"Using the discrete orthonormal basis, the ST of a discrete-time series x[kT] is expressed [23, 25] as and for n = 0, defined as where, j, m and n = 0, 1, 2, …, N - 1."

"The inverse DST is expressed as Consequently, other definitions of DST are also described in terms of discrete orthonormal ST (DOST) [22, 26], and expressed in terms of 'N' unit length vector, such as" where, k = 0, 1, 2, …, N - 1."

Methodology of the Image Encryption

"Algorithm: The image encryption technique is realized using the DOST with FrFT."

"The DOST can emphasize the highest energy of the base image on the minimum frequency sides."

"The image encryption method can be realized by expending FrFT."

"This specific technique is organized as i. Applying DOST on the original image A(x, y)gives the compressed form of a consistent lower frequency part B(u, v)."

Optical Realization of the Image Encryption

"The optical image encryption techniques for images have various advantages compared with standard methods like holographic electronic hybrid systems or realized in all-optical systems."

"The proposed convolution technique can be helpful for optical image encryption."

"It can serve as an efficient and undamaged way for image encryption."

"For image encryption constructed by the proposed method represents the original image to be encrypted, arranged as an intensity function, and the convolution of and represents the encrypted image."

"The simulation is used to test the reliability and security of the proposed technique."

Experimental Analysis and Results

"The superiority of decrypted images is verified by using MSE."

"The image of size M × N is expressed as where, Δx and Δy represents the pixel size."

"The peak signal to noise ratio can be defined as where, M × N is the size of the image."

Conclusion

"The DOST method is used because its computational complexity is less than the FFT method."

"We presented the image encryption algorithm using the concept of the DOST for the compression of the original image to a lower frequency component and then with different order and keys using the idea of FrFT."

"This transform is also used in other areas of signal processing and image processing like compression and image restoration."

"The decryption technique is a reverse process of the image encryption method."

Investigation of a Bridge Mechanical Response by a Joint Innovative Impulsive Energizer and a Wavelet Analysis [27]

This is a machine-generated summary of:

Caccamo, M. T.; Giuffrè, E.; Mezzatesta, F.; Pizzolo, F.; Tafaro, F.; Magazù, S.: Investigation of a Bridge Mechanical Response by a Joint Innovative Impulsive Energizer and a Wavelet Analysis [27]
Published in: International Journal of Civil Engineering (2021)
Link to original: https://doi.org/10.1007/s40999-021-00615-x

If you want to cite the papers, please refer to the original.
For technical reasons we could not place the page where the original quote is coming from.

Abstract-Summary
"An experimental approach to this issue requires to sufficiently energize the structure to allow the detection system to collect the entire range of the structure vibrational modes as well as to characterize their time evolution, also to get information on the structure degradation state."

"With the aim of monitoring the dynamical response of bridges, the present work addresses an approach based on both a wavelet transform and on an innovative cross-correlation wavelet analysis of the response signals obtained by a traffic-dependent energizer and by an impulsive energizer, where for this latter a new mathematical model has been formulated."

"It is shown that the analysis of a viaduct response, which also includes the cross-correlation wavelet method, makes possible to perform a joint time–frequency analysis for both the cases."

"The wavelet cross-correlation between the signals produced by the traffic-dependent energizer and by the impulsive energizer, furnishes a diachronic picture of its correlation coefficient."

Introduction

"The present work addresses an innovative approach based on a wavelet analysis of the response signal obtained by a traffic-dependent energizer, which corresponds to the normal working condition, and by an impulsive energizer [28–30] where the innovative aspect concerns with the application of the cross-correlation wavelet analysis [31, 32]."

"In both the cases, the aim is to characterize the vibrational features of the structure, through the acquisition and the spectral analysis of the response signal [33–35]."

"The wavelet analysis can be used for: (i) a time analysis of the frequency response of the selected structures; (ii) for a detail and approximation analysis of signals related to the dynamic response of systems; (iii) for signal denoising [1]; (iv) for detecting anomalies in the dynamic behavior of structures; (v) for the identification and determination of quantitative parameters relating to the state of deterioration of the structure [36–38].”

Wavelet Cross-Correlation Method

“FT describes the signal in terms of complex sinusoid while WT makes use of scaled and translated mother wavelet in a variable window.”

“The choice of the mother function is very fundamental to analyze the investigated signal and to choose the proper mother function the similarity of the shape and the bandwidth of the mother wavelet are necessary [39–41].”

“In the present study, we apply the Morlet mother function [42–44]: Furthermore, the cross-correlation wavelet analysis plays a crucial role in determining the degree of similarity, in the domain of time–frequency, between two signals thanks to cross-wavelet correlation (XWT).”

“To obtain detailed information about the delays in the oscillation, we calculate wavelet phase difference between the two signals: where < > indicate the average operator [45].”

“In the wavelet cross-spectrum, the phase is indicated by arrows.”

Experimental Set-Up and Measurements

“Our approach is based on a wavelet analysis of the response signal obtained by a traffic-dependent energizer, which corresponds to the normal working condition, and by an impulsive energizer.”

“The traffic-dependent energizer and the impulsive force allowed to sufficiently energize the structure modes and to identify the fundamental oscillation frequencies.”

“The vibration data acquisition was performed through the application of three acceleration sensors provided by PCB Piezotronics Inc.. These sensors were connected by cables to the Pimento multichannel analyzer and this latter was connected to a laptop.”

“A multichannel analyzer allowed to detect the acceleration values at the application points of the sensors.”

“After the signal denoising procedure, performed by means of the DWT we have evaluated both the Fourier and the continuous wavelet transforms (CWT) of the denoised signal.”

“To put into evidence the correlation between the signals produced by a traffic-dependent energizer and by the impulsive energizer, we performed a cross-wavelet correlation.”

Conclusions

"The present work addresses a dynamic analysis of mechanical structures, based on an innovative approach based on a wavelet analysis of the response signal obtained by a traffic-dependent energizer and by an impulsive energizer."

"The employed devices and approaches allow to realize the conditions for energizing the mechanical structure."

"It is shown that the analysis of a viaduct response through the wavelet method, makes possible to denoise the signal by means of DWT and to perform a joint time—frequency analysis showing for both the cases, traffic—dependent energizer and impulsive energizer, which frequencies are present and where in time they are, by means of CWT."

"The correlation between the signals produced by a traffic-dependent energizer and by the impulsive energizer was investigated by evaluating the cross-wavelet correlation coefficient of the two signals and its time evolution."

Low-Energy Real FFT Architectures and Their Applications to Seizure Prediction from EEG [46]

This is a machine-generated summary of:

Sanjeet, Sai; Sahoo, Bibhu Datta; Parhi, Keshab K.: Low-energy real FFT architectures and their applications to seizure prediction from EEG [46]

Published in: Analog Integrated Circuits and Signal Processing (2022)
Link to original: https://doi.org/10.1007/s10470-022-02094-z

If you want to cite the papers, please refer to the original.
For technical reasons we could not place the page where the original quote is coming from.

Abstract-Summary

"While many fast Fourier transform (FFT) architectures have been presented for computing real-valued FFT (RFFT), which of these architectures is best suited for low-throughput applications such as bio-medical signals which are typically sampled between 256 Hz and 1 kHz remains unclear."

"This paper implements and compares throughput, resources, and energy consumption of three different hardware architectures for real-valued FFT algorithms using Xilinx Ultra96-V2 FPGA development board."

"The RFFT architectures exploit the conjugate symmetry property of the real signals, thereby eliminating about half of the computations compared to a complex FFT."

"The seizure prediction system using the various RFFT architectures are then realized in Xilinx Ultra96-V2 FPGA development board and the power consumption values of the overall system using these architectures are compared."

Introduction

"The Cooley-Tukey algorithm assumes a general signal as the input to the FFT, i.e., the signal is assumed to be complex."

"Variants of the Cooley-Tukey algorithm have been proposed for real-valued signals [47–49], which reduce the complexity of the FFT even further."

"The architecture exploits the conjugate-symmetry property of the signal, thereby computing only 9 output samples as opposed to 16."

"While many FFT architectures have been presented for computing real-valued FFT (RFFT), which of these architectures is best suited for low-throughput applications such as biomedical where the signals are typically sampled between 256 Hz and 1 kHz remains unclear."

"This paper investigates different RFFT architectures in terms of hardware utilization, power consumption, and throughput for real-valued signals."

"The three FFT architectures investigated in this paper include: single processing element (SPE), pipelined, and in-place."

Seizure Prediction: An Overview

"As seizures are a result of abnormal electrical activities in the brain, Electroencephalogram (EEG) is an important tool for detection and prediction of seizures."

"Epileptic seizure prediction is done either in software or in hardware."

"There are two main stages for inference in seizure prediction: feature extraction, and classification of the extracted features into interictal or preictal."

"Seizure prediction is widely explored using supervised classification techniques, both in software and on hardware, with a majority of the methods using either time domain or frequency domain based feature extraction from EEG recordings."

"The proposed real-time seizure prediction system uses frequency content of the EEG signal extracted from its FFT as features for a Convolutional Neural Network based classifier."

"There have been a few works on implementing a seizure prediction system in hardware [50–56]."

"Yoo and others [54, 55] explored seizure prediction in hardware using linear and non-linear SVMs."

Proposed RFFT Architectures

"All the above FFT architectures process inputs and compute outputs in real time."

"A stage-counter tells which stage computation is being executed and the location-counter keeps track of the input to be sent to the PE and where the output needs to be stored."

"This is to ensure continuous computation, as one memory stores the output of the previous stage, and the other memory provides input for the computation of the output of next stage, thus facilitating real-time FFT computation."

"Similar to the SPE architecture, a stage counter keeps track of the stage under computation and appropriate logic is used for addressing the input and storing the output at the correct locations."

"Once the computation of the first stage completes, the second memory array acts as input, and the computed outputs are stored in the first memory array."

Seizure Prediction System

"The prediction engine processes in N-channel EEG signals of a patient as input."

"The frequency components over 30 s are sent to a CNN which classifies the EEG sequence as interictal or preictal."

"A total of 11 patients of the 23 are considered for training and testing the seizure prediction system."

"The feature used for classification is the frequency content of the EEG signal."

"The frequency information is extracted by taking the Short-Time Fourier Transform (STFT) of the signal."

"The feature map for classification is built for a signal duration of 30 s, resulting in 59 Fourier Transform output vectors per channel."

"This results in a total of 114 frequency components per Fourier transform per channel."

"A Convolutional Neural Network (CNN) is trained on the extracted feature map in a patient-specific way."

FPGA Implementation of RFFT Architectures

"The twiddle factors are also stored in LUTRAMs to facilitate the computation of two different outputs in a stage at the same time."

"Unlike the SPE architecture, the datapath in the fully-pipelined architecture is duplicated for the different stages of the FFT."

"The fully-pipelined architecture requires almost the same number of clock cycles for the computation of a single FFT as the SPE architecture, but 8 FFTs are computed in the pipeline."

"The throughput of the fully-pipelined implementation is 383142 FFTs per second when compared to the 47893 FFTs per second by the SPE architecture, both operating at a clock frequency of 100 MHz."

"The number of LUTs and FFs required by both SPE and pipelined architecture for 512-point FFT and 1024-point FFT would be 2.25 times and 5 times, respectively, as that of a 256-point FFT."

FPGA Implementation of Seizure Prediction System

"Since there is an overlap of 128 samples between consecutive FFTs, 23 FFTs have to finish computation before 128 input data cycles to ensure real-time computation."

"The input buffer is implemented as Block RAMs."

"Only a single instance of the FFT processor, the minimum operating frequency is calculated."

"The output of the FFT core consists of real and imaginary components of the frequency content separately."

"The design consideration is to achieve real-time operation with minimum hardware utilization."

"Since the operating frequency of the design is set to 100 kHz, 20, 240 clock cycles corresponds to 0.2 s, which is less than the constraint of 1 s."

"This implies that one processing element is sufficient to compute the first convolutional layer in real-time."

"One processing element takes 12, 800 clock cycles for the computation of this output and therefore is sufficient to ensure real-time processing."

Conclusion

"This paper implements and compares different hardware architectures for real-valued FFT algorithms."

"The architectures are based on a modified radix-2 FFT algorithm to utilize the fact that the Fourier transform of a real-valued signal is conjugate-symmetric."

"We conclude that the in-place architecture requires the least number of device resources when compared to the pipelined architecture, while the throughput of the pipelined architecture is approximately eight times that of the in-place architecture, for a 256-point FFT."

Quantitative Tissue Characterization Using Discrete Wavelet Transform of Photoacoustic Signals: A Feasibility Study [57]

This is a machine-generated summary of:

Rathi, Nikita; Sinha, Saugata: Quantitative tissue characterization using discrete wavelet transform of photoacoustic signals: a feasibility study [57]
Published in: Signal, Image and Video Processing (2019)
Link to original: https://doi.org/10.1007/s11760-019-01485-6

If you want to cite the papers, please refer to the original.
For technical reasons we could not place the page where the original quote is coming from.

Abstract-Summary
"This study introduces a method for extracting quantitative information about the size of regular-shaped PA absorbers using wavelet domain analysis of PA signals and concentration using segmentation."

"PA signals corresponding to tissue phantoms containing cylindrical PA absorbers were simulated under different conditions."

"The detail spectra of the simulated PA signals were computed using suitable mother wavelet and then further analysed to compute PA absorber diameter in terms of a distance parameter."

"A generalized method was proposed and verified using simulations for simultaneous quantitative recovery of absorber diameter and concentration for tissue phantoms containing PA absorbers with unknown radius and concentration values."

Introduction

"As the frequency spectrum of PA signal changes with the change in PA absorber shape and size, researchers employed frequency domain analysis of PA signal for tissue characterization."

"Although use of wavelet analysis is not new in the domain of PA imaging, only a few researchers used it for PA signal-based tissue characterization."

"We introduced a novel method for quantitative measurement of the dimension (diameter) of the regular-shaped (cylindrical) PA absorber using the detail wavelet spectrum of time-dependent PA signal generated by the absorber."

"Using detail simulations under different conditions, we showed that it is possible to recover quantitative information about diameter and concentration of PA absorbers using suitable analysis of PA signals."

"In similar studies performed earlier, researchers did not extract quantitative information related to PA absorber diameter and concentration simultaneously using PA signals."

Simulations for Computation of PA Absorber Diameter

"Our hypothesis is that detail spectrum, evaluated using a suitable mother wavelet, produces sharp peaks at the edges/sharp discontinuities in the PA signal and the distance between these sharp peaks provides the diameter of the PA absorber."

"Before evaluating detail wavelet spectrum of the simulated PA signal, we selected the suitable mother wavelet along with the proper decomposition level for computing the detail spectra of PA signals."

"To make the simulation close to actual soft tissue imaging, we considered that the PA absorber radius values follow Gaussian distribution with a specific mean and standard deviation."

"Detail spectrum of each simulated PA signal was computed and then distance between the major peaks in each detail spectrum was calculated."

"Similar to the simulation for initial PA signals, the total procedure was repeated for ten different Gaussian distributions with mean values 0.1 mm to 1 mm at an interval of 0.1 mm."

Simulations for Simultaneous Computation of PA Absorber Diameter and Concentration

"The radius values of the absorbers were assigned by sampling randomly from a Gaussian distribution of radius values with a specific mean and standard deviation while the distance values between consecutive PA absorbers were assigned by sampling randomly from a uniform distribution of distance values with specific maximum and minimum values."

"We computed time duration corresponding to signal segments (each segment contains PA signal generated by a single absorber) from the thresholded PA signal using zero crossing detection."

"We averaged the distance parameters from all the segments and computed mean diameter for corresponding distribution of PA absorbers constituting PA signal."

"As each segment corresponds to an individual PA absorber, counting the number of segments provided us concentration value."

"For all concentrations, PA absorber radii were sampled randomly from Gaussian distributions having same mean and standard deviations."

Discussion

"Since PA absorbers radii were sampled randomly from Gaussian distributions with same mean and standard deviation for all concentration values, the average recovered diameter versus actual concentration plot should ideally be a straight line with slope 0."

"In case of actual tissue imaging, the accuracy of the proposed method for quantitative recovery of PA absorber diameter and concentration will depend on the tissue structure, i.e. the size and distribution of the PA absorbers throughout the tissue."

"We will validate our findings of this simulation study using actual tissue phantoms containing PA absorbers of different shapes, size and concentration."

"The simulation results show that using the proposed method, it is possible to recover quantitative information about the PA absorber size from the detail spectrum of the PA signal generated by the cylindrical PA absorber and concentration by the segmentation."

Conclusion

"We investigated the feasibility of using discrete wavelet transform to analyse PA signals for quantitative tissue characterization through simulations on phantoms under different conditions."

"All the simulations were made close to the actual tissue imaging."

"A correlation between the size of PA absorbers and the parameter extracted from the detail wavelet spectra have been established which suggests that wavelet transform can potentially decode the quantitative information embedded in PA signals accurately and lead to better characterization of biological tissue."

An Efficient Short-Time Fourier Transform Algorithm for Grinding Wheel Condition Monitoring through Acoustic Emission [58]

This is a machine-generated summary of:

Lopes, Wenderson N.; Junior, Pedro O. C.; Aguiar, Paulo R.; Alexandre, Felipe A.; Dotto, Fábio R. L.; da Silva, Paulo Sérgio; Bianchi, Eduardo C.: An efficient short-time Fourier transform algorithm for grinding wheel condition monitoring through acoustic emission [58]

Published in: The International Journal of Advanced Manufacturing Technology (2021)
Link to original: https://doi.org/10.1007/s00170-020-06476-3

If you want to cite the papers, please refer to the original.

For technical reasons we could not place the page where the original quote is coming from.

Abstract-Summary

"AE signal analysis performed only in the frequency domain makes it impossible to locate faults on the grinding wheel surface during the dressing operation and examine the behavior of the frequencies contained in these signals over time."

"The time-frequency analysis of AE signals during dressing through STFT (short-time Fourier transform) can contribute toward the proposal of new monitoring methodologies, thus reflecting the optimization of the grinding process."

"Besides, this algorithm is used to investigate the characteristic frequencies in the aluminum oxide grinding wheel in dressing operation."

"The results indicate that the spectral content of the AE signals during dressing follows a uniform behavior, but their amplitude changes depending on the characteristics of topography and sharpness of the grinding wheel cutting edges."

Introduction

"This paper uses a groundbreaking approach, since a search in the databases failed to reveal any work that presents an algorithm based on the Kaiser window, which allows the STFT parameters to be adjusted to optimize the time-frequency resolutions when applied to the dressing operation."

"The STFT has been applied to the time-frequency analysis of signals acquired in the grinding process, such as in [59–67]."

"This research differs from others in the following aspects: In the literature, there were no studies that mention any proposed algorithm based on the Kaiser window and CCDM to determine the best possible time-frequency resolution for the STFT, and neither about the Kaiser window to compute the STFT of sensor signals acquired in the grinding process; Although the STFT was used in the time-frequency analysis of signals collected in the grinding process, the time-frequency analysis of AE signals in the dressing operation was not found in the literature."

State of the Art

"The authors concluded that the RMS of the AE signals and vibration above 10 kHz and 100 Hz, respectively, are related to the degradation of the grinding wheel's cutting surface and can be used as indirect methods to analyze grinding wheel service life."

"Lezanski [68] developed fuzzy logic and neural systems to classify the cutting conditions of an aluminum oxide grinding wheel in the cylindrical grinding of grade 45 carbon steel, using AE signals, vibration force and statistics derived from these signals as input data."

"The authors also showed that the kurtosis statistic applied to the RMS values of AE signals enabled them to analyze the progress of grinding wheel wear and to detect grinding burn on the machined surface."

"The studies developed by Moia and others [69] and by Nascimento Lopes and others [70], which aimed to optimize the dressing operation, proposed methods that classify grinding wheel wear and sharpness conditions using AE signals generated through experimental dressing tests."

Material and Methods

"The present work proposed an algorithm to set the best α to obtain the parameter β of the Kaiser window, which will provide the least background noise, leakage, and scattering, to obtain an appropriate time-frequency resolution to analyze the STFT spectrogram."

"Unlike existing approaches, in the present work, the CCDM index is computed for an appropriate frequency range and used as a representative metric to define the best parameter β to compute the STFT spectrogram of the EA signals coming from the dressing operation of aluminum oxide wheels."

"From a region of intersection of the average and standard deviation values of the α vectors obtained for the different conditions of the wheels, the best α was determined to obtain the value of β and then compute the STFT spectrogram of all signals of EA from both experimental tests 1 and 2."

Results and Discussion

"Resampling the signal at 500 kHz according to the Nyquist Theorem is justified because frequencies above this range, which have low energy levels, hinder the identification of a possible relationship of this harmonic content with the grinding wheel's conditions of topography and sharpness."

"For the dressed condition, it is observed that the AE signal presents more uniform amplitudes, which indicates that the cutting surface of the wheel is regular, i.e., full cutting capacity."

"Concerning the grinding wheel R100, it is observed in the average STFT spectra obtained for α equal to 21 dB, that the frequency with the highest amplitude level is approximately 55 kHz."

"The friction generated between the abrasive grains of grinding wheels R100 and R150 and the dresser's diamond tip produces greater acoustic activity, which generates AE signals with a higher and more uniform energy level along the dressing pass, and the harmonic content follows the same behavior."

Conclusions

"This paper proposes an algorithm based on the Kaiser window to adjust the STFT parameters so as to determine the best possible time-frequency resolution."

"The AE signals coming from the process were analyzed through the STFT, based on the parameters defined by the proposed algorithm."

"This enables the user to more easily define the appropriate parameters to calculate the STFT and thus maintain the balance of time-frequency resolution."

"The use of the algorithm to select the best parameters for calculating the STFT of the AE signals generated during dressing produced satisfactory results in the identification of the topographic conditions of the cutting surface of the grinding wheels under study."

"A comparison of the various windows revealed that the Kaiser window yielded the best results because it is easy to define the STFT parameters based on its set of equations, thereby achieving a balance between the time-frequency resolutions."

Deep Learning Classification of Motherboard Components by Leveraging EM Side-Channel Signals [71]

This is a machine-generated summary of:

Jorgensen, Erik J.; Werner, Frank T.; Prvulovic, Milos; Zajić, Alenka: Deep Learning Classification of Motherboard Components by Leveraging EM Side-Channel Signals [71]

Published in: Journal of Hardware and Systems Security (2021)
Link to original: https://doi.org/10.1007/s41635-021-00116-2

If you want to cite the papers, please refer to the original.
For technical reasons we could not place the page where the original quote is coming from.

Abstract-Summary
"We show that a broad range of motherboard components can be classified based on their electromagnetic (EM) emanations using a lightweight convolutional neural network (CNN) architecture."

"Our random bootstrap sampling approach to cross-validation allows us to select robust models that achieve strong accuracy when classifying processor, memory, power, and ethernet integrated circuits (ICs) on internet of things (IoT) devices in two different excitation states."

"An analysis of signals passing through our end-to-end trained model allows us to generate insights about the discriminative features in the emanations of different components."

"Our results demonstrate that a CNN architecture can be used to classify ICs accurately using a diverse set of EM emanations while providing insights about the relevant discriminative features between components."

Introduction

"The production pipeline of any electronic device involves multiple manufacturers and suppliers, each of which could introduce counterfeit components to a device."

"Electrical inspection may involve testing components with burn-in tests, though code still may execute normally on counterfeit chips and make it difficult to find security flaws without more exhaustive and time-consuming testing [72]."

"While these EM side-channel techniques are successful in classifying whole devices, they have not been shown to identify individual ICs or components within those devices."

"There is no established state-of-the-art solution to the problem of utilizing EM side channels to accurately identify individual components on an assembled motherboard."

"This classification is important for device designers, who need ways of validating that the components on their devices have not been replaced by a third-party assembler or supplier with counterfeits, such as a cheap alternative or inauthentic copy."

"We present a generic approach for accurate and interpretable component-level classification and counterfeit detection on IoT devices."

Threat Model

"Upon receiving a fabricated device, the designer can validate the authenticity of components by measuring their EM side-channel emanations and feeding the signal through the model presented in this work."

"Each third-party supplier or assembler in the device manufacturing pipeline introduces a potential vulnerability to counterfeit components being swapped in for their authentic counterparts."

"Component validation is most simple and reliable for the device designer to perform when the device is powered on in an idle state without loaded program activity."

"The device designer has knowledge of at least one component in addition to the authentic component that could be a potential counterfeit swapped in for the original."

"The neural network model we present here extracts discriminating features from the EM side-channel emanations of a known set of components and devices."

"The feature embeddings learned by the proposed CNN are used to differentiate between known component models that could be swapped in as counterfeits."

Structured Signals

"With repeated excitation, such as a device clock signal or repeating code pattern, we expect to see emanations that are especially noticeable in the frequency domain."

"We measure emanations when the device is in an idle state and also when it is activated by a controlled, repeating code execution."

"We execute a repeating code pattern to generate amplitude modulation with the clock frequency [73]."

"Distinguishing different component models in the excited state, the same features may be used in addition to the alternating frequency, relative excitation and harmonic magnitudes, and modulations on other unknown sources."

Preprocessing and Model Architecture

"Each EM emanation measurement can contain millions of samples due to the large bandwidth needed to capture the alternating frequency and harmonics of excited components."

"We subsample each measurement to increase the number of training samples available, reduce the dimensionality of the inputs, and capture potential random time-variability of the signal."

"When measuring EM emanations while executing repeating code patterns, we expect signal structure to be revealed best in the frequency domain, so each subsample is converted to the frequency domain with a Short-Time Fourier Transform (STFT)."

"The result of the strided convolution and pooling is a downsampling of twelve times in the length of samples fed through the layer."

"The first convolution block takes as input a univariate subsample of length 4096 and learns three single-channel filters."

"The five-channel output of the convolution layers is flattened to a single vector of length 140 to be fed to the linear layers."

Experimental Design

"The model that achieves the best validation accuracy across the R different random training/validation combinations is saved and evaluated on the test set to measure its overall performance."

"We perform these tests on memory components since we have the greatest diversity of memory components across device types and training our CNN requires at least two classes to build class-discriminating features."

"For each of our five tests, we train the CNN with one memory component model held out of the training and validation sets."

"The CNN is then trained with only four output neurons now representing the "known" four memory component models."

"We have nine or ten copies of each device type, resulting in a total of 273 distinct components measured across the processor, memory, power, and ethernet models with which we employ a 10%/30%/60% training/validation/testing set split."

Results and Interpretability

"Prediction accuracies here correspond to the exact fraction of components predicted correctly in the test set for each component class."

"A memory classification accuracy of 0.952 corresponds to 60 of the 63 test set components being predicted correctly with three predicted incorrectly."

"The distinct emanation signatures of the different power management ICs in the idle state yield much better prediction accuracy."

"Our method yields the lowest accuracy when detecting an anomalous memory IC model which has EM emanations similar to a model with which the CNN was trained."

"Although the inputs for each class look similar, we note subtle differences in the height of the carrier peak, presence of small modulated excitations, slope of the sidebands, and sharpness of the trail-off at the outer edges of the measured bandwidth."

Conclusion

"Our work develops a lightweight CNN to classify a broad range of motherboard components in idle and excited states based on their EM emanations."

"Our robust preprocessing and cross-validation pipeline allows us to train models with strong classification accuracy across a range of component classes."

"We show that a simple anomaly detection procedure can be used with our CNN to detect counterfeit ICs on which the CNN was not trained."

"While our network performs well across a variety of component classification problems, there is room for improvement with future work."

"Our results demonstrate that a lightweight CNN can classify diverse EM emanations accurately while generating insights about their discriminative features."

Recent Advances in the Application of Deep Learning for Fault Diagnosis of Rotating Machinery Using Vibration Signals [74]

This is a machine-generated summary of:

Tama, Bayu Adhi; Vania, Malinda; Lee, Seungchul; Lim, Sunghoon: Recent advances in the application of deep learning for fault diagnosis of rotating machinery using vibration signals [74]
Published in: Artificial Intelligence Review (2022)
Link to original: https://doi.org/10.1007/s10462-022-10293-3

If you want to cite the papers, please refer to the original.

For technical reasons we could not place the page where the original quote is coming from.

Abstract-Summary

"Vibration measurement and monitoring are essential in a wide variety of applications."

"Vibration measurements are critical for diagnosing industrial machinery malfunctions because they provide information about the condition of the rotating equipment."

"Numerous studies conducted in this vein have been published in a variety of outlets."

"This review documents data-driven and recently published deep learning techniques for vibration-based condition monitoring."

"The selected studies are then systematically discussed to provide researchers with an in-depth view of deep learning-based fault diagnosis methods based on vibration signals."

Introduction

"For machines that work under different speed conditions, the non-stationary vibration signals [75, 76] can be preprocessed by using the speed signals so that the preprocessed signals can be readily employed for diagnosis."

"A variety of techniques for obtaining the speed signal from vibration measurements have been studied, with the vast majority relying on the time-frequency (TF) representation technique [77–80]."

"This paper focuses on documenting an overview of DL techniques for condition monitoring using vibration data."

"This review article is an extension of a previously published paper in Tama and others [81] and has two main contributions: (i) To provide documentation on how DL techniques have been employed and are presently addressed for vibration signal-based machine fault diagnosis; (ii) To present a benchmark on which DL models from the selected studies have performed better."

Fundamental Concepts

"Raw signals obtained from rotating machinery are typically in the time domain, representing a collection of time-indexed data points (e.g., acceleration, velocity, and background noise) obtained over a period of time."

"CNN, a type of neural network that specializes in processing data with a grid-like topology, was first introduced in the LeNet-5 architecture [82] and is referred to as convolutional 2D. CNN is typically utilized on image data."

"Four fundamental phases can be described as follows. (i) Graphs may express the geometry and structure of data as input to a neural network and therefore provide more information than other general data formats. (ii) Fast-localized filters based on graph spectral theory and Chebyshev expansion are utilized to conduct a convolution operator on a graph. (iii) A graph coarsening process is conducted to group similar vertices together. (iv) For graph pooling, the vertices are rearranged by first generating a balanced binary tree then applying regular 1D pooling."

Methodology

"We searched two well-known indexing databases, such as Scopus and Web of Science, to incorporate all prior works comprehensively."

"We limited our search to these databases because: (i) the databases are supposed to archive high-quality papers; and (ii) to avoid multiple searches, which are typically available at each publication outlet."

"Some search keywords were derived from index terms found in some papers."

"All studies considered in this review must be papers that were published in journals because journal articles represent the highest level of research outcomes and have been thoroughly peer-reviewed [83]."

Result and Discussion

"The authors in [84] propose a method called dislocated time series CNN to develop a new diagnosis framework based on the characteristics of industrial vibration signals."

"Another CNN variant named deep capsule network with a stochastic delta rule (DCN-SDR) for rolling bearing fault diagnosis is proposed in [85]."

"A study presented by [86] discusses a fault-bearing diagnosis method based on RNN in the form of an autoencoder."

"The proposed method shows robust and superior performance for the two datasets, including cutting tool monitoring and bearing fault diagnosis."

"To handle the unbalanced dataset problem for fault diagnosis, Liu and others [87] use an autoencoder-based SuperGraph feature learning method."

"To handle data limitations, Zhao and others [88] propose a new semisupervised deep convolutional belief network (SSD-CDBN) for motor-bearing fault diagnosis."

Future Research Directions

"This current study has reviewed the application of DL algorithms for fault diagnosis methods using vibration signals."

"The interpretability of DL algorithms for fault diagnosis utilizing vibration signals is still underexplored."

"While the DL algorithm for fault diagnosis is a solely data-driven algorithm that may well suit observations, predictions may be inconsistent or unreasonable due to extrapolation or observational biases, resulting in poor generalization performance."

"There is an urgent need to incorporate fundamental physical principles and domain knowledge into the DL algorithm in order to govern physical norms, which could provide informative priors or substantial theoretical limitations and inductive biases on top of observational biases."

"Although the transformer is a revolutionary design for DL model architectures, it has been underexplored in the realm of fault diagnosis until Pei and others [89] introduced the transformer convolution network (TCN) for rotating machinery fault diagnosis."

Threat to Validity

"Some search keywords were drawn from keywords found in some papers to reduce the risk of excluding relevant studies."

"The search was restricted to two primary indexing databases to minimize bias and miss relevant papers."

"A consensus mechanism was utilized to select all the extracted studies to minimize this issue."

Conclusion

"DL techniques have solved complex problems across various applications and research concerning the fault diagnosis of rotating machinery."

"Some current research challenges were reported, providing researchers insights into the latest DL techniques for fault diagnosis."

"A 'white-box' model for fault diagnosis will be sought after as it helps the prediction be understandable to humans."

"This study has observed DL for fault diagnosis using vibration signals either in supervised, unsupervised, or semi-supervised learning mechanisms."

A Novel Intelligent Approach Based on WOAGWO-VMD and MPA-LSSVM for Diagnosis of Bearing Faults [90]

This is a machine-generated summary of:

Taibi, Ahmed; Ikhlef, Nabil; Touati, Said: A novel intelligent approach based on WOAGWO-VMD and MPA-LSSVM for diagnosis of bearing faults [90]

Published in: The International Journal of Advanced Manufacturing Technology (2022)
Link to original: https://doi.org/10.1007/s00170-022-08852-7

If you want to cite the papers, please refer to the original.

For technical reasons we could not place the page where the original quote is coming from.

Abstract-Summary

"The vibration signals of rolling bearings are often nonlinear and nonstationary; it is difficult to extract sensitive features and diagnose faults by traditional signal processing methods."

"A novel intelligent fault-diagnosis approach based on whale optimization algorithm grey wolf optimization-variational mode decomposition (WOAGWO-VMD) algorithm and the marine predators algorithm optimization-least squares support vector machine (MPA-LSSVM) is proposed in this paper."

"The optimized VMD algorithm is utilized to decompose the vibration signal of the rolling bearing into several intrinsic mode functions, and a new sensitive indicator is created to select the components containing the most information."

"The feature vectors obtained are input to the MPA-LSSVM for diagnosis and identification."

"The results have shown that the proposed method can extract the fault feature information of 16 bearing signals of different fault types effectively and identify them accurately."

Introduction

“Fu and others [91] proposed a new hybrid approach, coupling VMD algorithm, composite multiscale fine-sorted dispersion entropy (CMFSDE), and support vector machine (SVM) for fault diagnosis of rolling bearings.”

“Jin and others [92] proposed a fault diagnosis method of train rolling bearing based on VMD to decompose the original signal into several intrinsic mode components (IMF) and calculated the distribution entropy of each component as the feature vector.”

“Cheng and others [93] proposed a bearing fault diagnosis method based on VMD and singular value feature (SVD) for feature extraction and G-G fuzzy clustering for fault identification.”

“We consider in this paper the multi-features based on dispersion entropy feature (DE), permutation entropy feature (PE), and SVD as the fault feature vector of rolling bearing vibration signal.”

“We used the marine predator algorithm MPA to optimize the parameters of classifier LSSVM; hence, the optimized model is used to classify and identify the different fault types of rolling bearing.”

Fault Feature Extraction

“To solve this limitation, we introduce an algorithm called WOAGWO, proposed by Mohammed H [94]; this algorithm is a hybrid between the WOA and GWO algorithm in order to improve the performances of WOA and to obtain better solutions.”

“Pseudocodes of WOAGWO algorithm: It is noteworthy that the parameter settings [k, α] of VMD need to be set in advance when decomposing signals due to their high impact on results [95], where alpha is directly proportional to bandwidth since smaller alpha gives a smaller bandwidth and vice versa; on the other hand, smaller K will give raise to mode aliasing and if K is too large it will lead to useless component generation [96].”

“In order to validate the effectiveness of the WOAGWO-VMD algorithm proposed in this paper, it is applied to decompose the simulation signal mentioned in the literature [97].”

“The optimized VMD algorithm with the optimal combination [K, α] is used to decompose the simulation signal.”

“Obtained, it can be concluded that the proposed WOAGWO-VMD algorithm can decompose the frequency components in the simulation signal successfully, and overcome the mode mixing phenomenon in EMD algorithm.”

Fault Feature Classification Using MPA-LSSVM

"The detailed steps of the algorithm are presented as follows: This is the most important process in the initial iteration of improvement where exploration is important; at this point, when the movement of the predator is faster than the prey, the best strategy for the predator is to not move."

"Computational complexity of MPA algorithm is O(t(nd + cof × n)), where cof is the evaluation function, t is the iteration number, d is the optimization problem dimension, and n is the number of populations."

"To improve the classification feasibility of LSSVM, an algorithm should be used to optimize the parameters (c and g)."

"Step 3: Use the following equation as fitness function of MPA-LSSVM."

"The aim of the LSSVM parameter optimization problem is to minimize the fitness function."

"Step 6: Export the optimal values of (c_{best}, g_{best}) once the maximum number of iterations is reached and trained LSSVM model."

Fault Diagnosis Method Based on WOAGWO-VMD and MPA-LSSVM

"The specific steps of this method are summarized as follows: Step 1: The bearing vibration acceleration signals in various states (i.e., normal, ball fault, inner race fault and outer race fault) are collected using the acceleration sensors."

"Step 3: Utilize VMD with optimal parameter combination [K_0, α_0] to decompose the vibration signal into several IMFs."

"Step 5: Select four IMF components with greater correlation with the original signal, and calculate DE, PE, and SVD of each IMFs, then construct the multiple features vector."

"Step 7: The obtained feature vectors normalized are randomly divided into two groups, a training sample set and a testing sample set."

Experimental Verification and Result Analysis

"To achieve the intelligent bearing fault diagnosis, the matrix of multi-features vector obtained were input to the MPA-LSSVM classifier for fault classification and recognition."

"The diagnosis effect of using the optimized VMD is better than that of the unoptimized VMD method, which indicates the optimized VMD can more accurately extract the fault feature information of the rolling bearing."

"The proposed method achieved a higher classification accuracy in the diagnosis of various types of faults than the other methods."

"In order to further verify the effectiveness and superiority of the proposed method for diagnosis of bearing faults, we verify the superiority of our method used for feature extraction (WOAGWO-VMD decomposition and multi-features) as first approach by using the fault classifiers ELM, LSSVM, PSO-LSSVM, and PSO-SVM applied in the literature He et al. [98], Pratama et al. [6], Li et al. [99], and Liang et al. [100] respectively to diagnose and identify 16 bearing signals."

Conclusions

"This paper presented a novel intelligent rolling bearing fault diagnosis method based on WOAGWO-VMD and MPA-LSSVM."

"The WOAGWO-VMD algorithm and multi-features were used for fault feature extraction, and the MPA-LSSVM algorithm for fault classification."

"Through the simulated signal that was analyzed, the WOAGWO-VMD algorithm can extract the fault feature information of the signal more powerfully and significantly compared to EMD, which is critical for the diagnosis of rolling bearing faults."

"The effectiveness and feasibility of the proposed method in this paper is verified by using experimental data of rolling bearing fault vibration signal."

Fault Diagnosis in Bevel Gearbox Using Coiflet Wavelet and Fault Classification Based on ANN Including DNN [101]

This is a machine-generated summary of:

Babu, T. Narendiranath; Ali, P. Sahir Nowshad; Prabha, D. Rama; Mohammed, V. Noor; Wahab, Razia Sultana; Vijayalakshmi, S.: Fault Diagnosis in Bevel Gearbox Using Coiflet Wavelet and Fault Classification Based on ANN Including DNN [101]

Published in: Arabian Journal for Science and Engineering (2022)

Link to original: https://doi.org/10.1007/s13369-022-06767-9

If you want to cite the papers, please refer to the original.

For technical reasons we could not place the page where the original quote is coming from.

Abstract-Summary
"It is very important to detect the faults in the machinery as early as possible in order to stop the propagation of faults which may lead to heavy damages."

"The objective of our study is to investigate the vibration analysis of bevel gearbox to detect the faults."

"Various types of faults were induced in the gearbox."

"The vibration analysis was carried out for different types of faults with different lubrication levels in the gearbox."

"An artificial neural network (ANN) and a deep neural network (DNN) were used to detect the faults in the bevel gearbox automatically."

Introduction

"Sharma and others [102] worked on the fault detection in a bevel gearbox using discrete wavelet transform and autoregressive model."

"An autoregressive model was built using coefficients of the discrete wavelet transform to detect the faults in the gearbox."

"Huang and others [103] studied the use of wavelet packet transform to extract fault features from the signals and the RS for classification."

"Li and others [104] proposed a method to detect crack faults in a gearbox using discrete wavelet transform and an adaptive resonance theory neural network."

"Wang and others [105] used wavelet transform for the fault diagnosis of the gearbox."

"Chen and others [106] proposed a method based on discrete wavelet transformation and convolution neural network to identify the fault conditions of planetary gearbox in wind turbines."

"The vibration signal was decomposed adaptively using empirical wavelet transform for the identification of compound faults in a gearbox."

Experimental Setup and Procedure

"The experimental setup consists of a spiral bevel gearbox with a gear ratio of 2.5."

"The vibration signals were acquired in the time domain at the sample rate of 20 kHz for 10 s with full (600 ml), half (300 ml) and quarter (150 ml) lubrication levels in the gearbox using LabView software."

"These types of faults occur due to continuous running of gearbox for a long period of time."

"The lubrication oil generally gets contaminated due to the debris of teeth cuts, edge faults, etc , and also when the gearbox is left open during assembly."

Gear Mesh Frequency (GMF)

"Gear mesh frequency and the sidebands around it are important in the fault diagnosis."

"Increase in the number of sidebands spaced around the GMF and increase in the amplitude of the sidebands represent the presence of a problem."

"This results in spread of few number of sidebands around GMF and its harmonics over a range of frequency."

"Due to vibration and noise from other mechanical components, the detection of presence and spacing of sidebands around GMF in the frequency spectrum is difficult."

Signal Processing

"The wavelet transform is used as a filter bank to denoise the signal."

"Wavelet transform has replaced conventional filters in many applications due to its ability to extract information in both frequency and time domains from the transient signals."

"The wavelet transform has a great advantage of denoising the signal as its purpose is to de-noise the signal by decomposing and reconstructing it into original form without degradation of information."

"Based on these advantages, the Coiflet 1 wavelet was preferred to denoise the signal."

"The denoising capability of the wavelets was evaluated using three parameters, namely mean square error (MSE), signal-to-noise ratio (SNR) and peak signal-to-noise ratio (PSNR)."

"As the signal denoised with Coiflet wavelet gave the least MSE value and higher SNR and PSNR values, it can be concluded that the Coiflet wavelet has the better denoising capability when compared to biorthogonal, Daubechies and Symlet wavelets."

Artificial Neural Network (ANN)

"The neural network used in this work was a feedforward network with sigmoid hidden neurons and softmax output neurons."

"A feedforward neural network is a classification algorithm where the connections present between the neurons do not form a cycle."

"Once the data enter the input layer, they reach the output layer passing through the network, layer by layer."

"Overfitting is the case where the capacity of the network to process the information is very high but the limited data is not sufficient to train each and every neuron present in the hidden layers [107]."

"For the ANN, the data were trained using the neural network with a single hidden layer."

"These statistical features were given as input to train the neural network."

"The neural network is trained by dividing the input data into three different subsets."

Deep Neural Network (DNN)

"Deep neural network is an artificial neural network with multiple hidden layers to make it understand very complex patterns."

"When compared to a two-layered neural network, the deep neural network solves the problem more efficiently and make predictions as well as conclusions depending on the inputs given to the network and desired results."

"The deep neural network was used to train the network to predict the desired outputs with high accuracy."

"The deep neural network contains multiple hidden layers."

"Autoencoders are a type of feedforward neural networks whose outputs are very identical to the inputs."

"There are various reasons to use autoencoder in the deep neural network."

"The deep neural network was trained using scaled conjugate gradient and the cross-entropy error was used to analyse the performance of the network, as already discussed in the previous section."

Results and Discussion

"It can be seen from the plots of filtered signals of gearbox with three different levels of lubrication that the amplitude of GMF is increasing as the number of faults are increasing in the gearbox."

"The number of epochs parameter will be set before training the network, but the training of the network stops when the validation error starts increasing."

"The training of network was stopped when the validation error increased consecutively for six epochs which was set as a default in MATLAB."

"The training state plot shows that the validation checks is equal to zero at 1000 epochs which suggests that the error never failed to decrease during the entire training of the network."

"The accuracy has increased due to increase in the number hidden layers which allows the neural network to perform deeper analysis to recognize the minute changes in the data to identify the faults in the gearbox."

Conclusions

"The results of the experiment are validated as well as the adaptability of the wavelet network and the gearbox vibration data has been verified using the frequency plots by comparing the calculated theoretical GMF with actual GMF obtained from the frequency plots, which are approximately same."

"The artificial neural network was useful for automatic fault detection."

"The deep neural network was used to predict the data more accurately as the deep neural network is necessary when the data are overlapped due to more number of fault conditions."

"The accuracy of the deep neural network is 100% which means that the network has classified all the faults accurately."

Machine Learning-Based EEG Signals Classification Model for Epileptic Seizure Detection [108]

This is a machine-generated summary of:

Aayesha; Qureshi, Muhammad Bilal; Afzaal, Muhammad; Qureshi, Muhammad Shuaib; Fayaz, Muhammad: Machine learning-based EEG signals classification model for epileptic seizure detection [108]

Published in: Multimedia Tools and Applications (2021)
Link to original: https://doi.org/10.1007/s11042-021-10597-6

If you want to cite the papers, please refer to the original.

For technical reasons we could not place the page where the original quote is coming from.

Abstract-Summary

"The detection of epileptic seizures by classifying electroencephalography (EEG) signals into ictal and interictal classes is a demanding challenge, because it identifies the seizure and seizure-free states of an epileptic patient."

"Several machine learning-based strategies were introduced to investigate and interpret EEG signals for the purpose of their accurate classification."

"In order to address this issue, this paper focuses on extracting the most discriminating and distinguishing features of seizure EEG recordings to develop an approach that employs both fuzzy-based and traditional machine learning algorithms for epileptic seizure detection."

"The proposed framework classifies unknown EEG signal segments into ictal and interictal classes."

Introduction

"The trained algorithms build classifying models that perform the classification of EEG signals for the automatic detection of epileptic seizures."

"The available approaches of automatic epileptic seizure detection encounter some challenges that lead to the inaccurate classification of EEG signals."

"One major challenge is the extraction of the most representative and distinguishing features of epileptic EEG signals."

"Since the dynamic nature of EEG signals causes variations in the characteristics of seizure recordings acquired from different seizure events, there are complications in extracting the most discriminating feature set of epileptic EEG signals."

"To improve the accurate detection of epileptic seizures, fuzzy-based and traditional machine learning algorithms are employed for the classification of EEG signals."

"An improved machine learning-based epileptic seizure detection model is proposed that categorizes EEG signals into ictal and interictal classes with an enhanced accuracy percentage."

Previous Approaches for Epileptic Seizure Detection

"The current methods use different signal processing techniques to extract useful features for classifying EEG signal recordings."

"The characterizing features of decomposed signal sub-bands are extracted and input to machine learning algorithms that build models for classifying EEG signals into normal, ictal, and interictal classes."

"The extracted temporal features built an SVM classification model that showed high false detection rates for clinical EEG and a low sensitivity of 94.5% for behind the ear EEG."

"In [109], A. Bhattacharyya and R. B. Pachori proposed a framework that built an RF-based classification model using the features of EWT decomposed EEG signals."

"L. S. Vidyaratne and K. M. Iftekharuddin [110] extracted HWPT decomposed EEG signals and non-linear features to build a classification model of RVM."

"The extracted features built a novel RF classifier to perform classification and then identification of epileptic EEG signals."

Proposed Model

"It consists of six major steps: (1) EEG dataset collection, which describes the collected EEG signal recordings; (2) pre-processing, which removes artifacts from the input dataset; (3) signal splitting, which splits long EEG recordings into small epochs; (4) feature extraction, which extracts temporal and spectral features from decomposed signal sub-bands; (5) classifier building, which builds a classifier using traditional and fuzzy machine learning algorithms on the extracted feature vector; and (6) classifier evaluation, where the constructed classification models are evaluated using accuracy, sensitivity, specificity, precision, and the Matthews Correlation Coefficient (MCC)."

"Five traditional machine learning algorithms, K-Nearest Neighbor (KNN), Decision Tree (DT), Random Forest (RF), Support Vector Machine (SVM), and Multilayer perceptron (MLP), and five fuzzy machine learning algorithms, Fuzzy Unordered Rule Induction Algorithm (FURIA), Fuzzy Nearest Neighbor (FNN), Fuzzy Rough Nearest Neighbor (FRNN), Vaguely Quantified Nearest Neighbor (VQNN), and Fuzzy Lattice Reasoning (FLR), are applied to build models for EEG signal classification."

Experiments and Results Discussion

"In the first experimental set, the single-channel Bonn EEG dataset is used for evaluation, while in the second experimental set, the multi-channel CHB-MIT EEG dataset is utilized along with five traditional algorithms (RF, SVM, KNN, DT, MLP) and five fuzzy machine learning algorithms (FURIA, FNN, FRNN, VQNN, FLR)."

"This section discusses an analysis of the results obtained by employing traditional and fuzzy classifiers using the single-channel EEG recordings of the Bonn dataset."

"In these results, FRNN showed the best performance, with 99.81% accuracy and 100% specificity."

"In the case of the fuzzy classifiers, FRNN provided the maximum performance for all metrics, with 98.70% accuracy."

"Among other classifiers, VQNN achieved an accuracy of 97.59% and a specificity of 97.46%, but FNN yielded 96.85% accuracy with 98.84% sensitivity."

"The classification results of traditional and fuzzy classifiers are analyzed using the multi-channel scalp EEG recordings of the CHB-MIT dataset."

Comparative Analysis

"In NS-S classification case of CHB-MIT dataset LSTM achieved the highest accuracy of 88.76% while the best specificity of 90.83% provided by DNN."

"By using HyperGCN, the best scores as 96.63% accuracy with 96% sensitivity, 0.967 precision, and 0.595 mcc obtained in C-E classification case, however, the best specificity of 97.92% was resulted while performing CD-E classification case of Bonn dataset."

"In the case of the Bonn dataset, [111–115] provided percentage accuracy of 99%, 98.5%, 99%, 99.54%, and 99.5% respectively for a C-E classification case that improved to 99.81% by the proposed method."

"Using the CHB-MIT dataset, the proposed method showed an improved classification accuracy of 92.79% as compared to [116, 117], which provided 90% and 85.6% accurate classification of EEG signals for the NS-S classification case respectively."

Conclusion

"The detection of epileptic seizures using machine learning-based approaches for EEG signal classification has been frequently employed in the literature."

"A feature extraction method was proposed that inspected the most discriminating features of EEG signals to perform an effective classification."

"Fuzzy and traditional machine learning algorithms were employed for the classification of EEG epochs into interictal and ictal classes."

"As compared to the scalp EEG from the CHB-MIT dataset, intracranial EEG recordings from the Bonn dataset provided improved scores in all performance measures for classification between interictal and ictal classes."

Artificial Intelligence Methods for Analysis of Electrocardiogram Signals for Cardiac Abnormalities: State-of-the-Art and Future Challenges [118]

This is a machine-generated summary of:

Saini, Sanjeev Kumar; Gupta, Rashmi: Artificial intelligence methods for analysis of electrocardiogram signals for cardiac abnormalities: state-of-the-art and future challenges [118]

Published in: Artificial Intelligence Review (2021)
Link to original: https://doi.org/10.1007/s10462-021-09999-7

If you want to cite the papers, please refer to the original.
For technical reasons we could not place the page where the original quote is coming from.

Abstract-Summary
"Early detection of cardiac arrhythmias through automatic electronic techniques is an important research field since the invention of electrocardiogram (ECG or EKG) and advanced machine learning algorithms."

"ECG (EKG) provides the record of variations in electrical activity associated with the cardiac cycle, used by cardiologists and researchers as a gold standard to study the heart function."

"The present work is aimed to provide an extensive survey of work done by researchers in the area of automated ECG analysis and classification of regular & irregular classes of heartbeats by conventional and modern artificial intelligence (AI) methods."

"The artificial intelligence (AI) based methods have emerged popularly during the last decade for the automatic and early diagnosis of clinical symptoms of arrhythmias."

"The literature is explored for the last two decades to review the performance of AI and other computer-based techniques to analyze the ECG signals for the prediction of cardiac (heart rhythm) disorders."

"The existing ECG feature extraction techniques and machine learning (ML) methods used for ECG signal analysis and classification are compared using the performance metrics like specificity, sensitivity, accuracy, positive predictivity value, etc Some popular AI methods, which include, artificial neural networks (ANN), Fuzzy logic systems, and other machine learning algorithms (support vector machines (SVM), k-nearest neighbor (KNN), etc) are considered in this review work for the applications of cardiac arrhythmia classification."

"The aim is to provide the reader, the prerequisites, the methods used in the last two decades, and the systematic approach, all at one place to further purse a research work in the area of cardiovascular abnormalities detection using the ECG signal."

Introduction

"Any abnormality in the operation of heart gives rise to Cardiovascular Diseases (CVDs) including heart attacks, strokes, and cardiac arrhythmias."

"Researchers have also used artificial neural networks for feature extraction and classification of ECG signals for remote monitoring and analysis of heart diseases."

"Other than the irregularities in cardiac rhythm, myocardial infarction (i.e. heart attack) is a major heart disease that can result in the sudden death of a patient due to the lack of oxygenated blood supply to the heart muscles."

"Researchers have developed some automatic methods with the aid of artificial intelligence (AI) methods for the early detection of ST-segment elevation myocardial infarction (STEMI) using electrocardiogram (ECG) signals in order to reduce the risk of STEMI."

Electrocardiogram (ECG) and Cardiac Cycle

"The difference in bio-potentials captured by electrodes from the body surface is a weak electrical signal which is processed through the different stages of a signal conditioning circuit to get an ECG record in digital form [119]."

"These capacitive-coupled sensors record the variations in electrostatic charge associated with muscle activity occurring in the heart cycle similar to the standard ECG signal."

"To capture the bio-potentials associated with heart activity, propagating at the body surface, electrodes (sensors) are placed at different points of the body (chest and limbs), which are connected through wires and instrumentation to the processing and recording device such as an electrocardiograph."

"The traditional ECG cardiographs utilize a standard 12 lead system (through 10 electrodes) where each lead records the voltage signal generated by cardiac activity from different angles to completely characterize the heart's electrical activity graphically [119–121]."

ECG Feature and Extraction Methods

"As electrocardiogram signals are non-stationary, researchers have applied various time–frequency domain feature extraction methods to reveal the time-varying morphology of ECG signals."

"ECG features can be extracted in time, frequency, and time–frequency domain using different mathematical tools available for signal analysis."

"Feature extraction from ECG signal in time-domain has been done by researchers using LPC method by calculating the LPC model coefficients from the selected time durations of the signal and combining them as an input feature vector for the classification stage [122–124]."

"Like the other real world signals, ECG signals are non-stationary in nature (changing morphology with respect to time) and frequency domain features (extracted fro DFT/FFT) alone are not sufficient to provide the localized frequency component information in the signal at different time instant [125, 126]."

Heartbeat/Arrhythmia Classification

"The popular artificial intelligence (AI) methods available for analysis and classification of ECG signals are: Artificial Neural networks Fuzzy Systems Neuro-Fuzzy systems Genetic-Fuzzy systems Probabilistic Neural Networks (PNN) Convolutional Neural Networks (CNN) Support Vector Machines (SVMs) Linear Discriminants (LDs) To avoid the conventional, time consuming, and complex methods for hand crafted feature extraction of biological signals that requires expert domain knowledge, deep convolutional neural networks (CNN) are becoming popular for the convinient feature extarction and classification methods of such time-based signals."

"Pourbabaee and others [127] suggested a deep learning method based on convolution neural network (CNN) for feature extraction and classification of ECG signals for the automatic diagnosis of paroxysmal atrial fibrillation (PAF)."

"Raj and Ray [128] also presented a PSO optimized support vector machine classification method using time–frequency features of ECG derived using discrete cosine transform (DCT) for classifying 16 different arrhythmia classes on MIT-BIH arrhythmia database."

Conclusion

"The study reveals that researchers have applied different artificial intelligence techniques and machine learning algorithms on the different sets of features extracted from the ECG signals to categorize the heartbeats in different classes of arrhythmia and normal sinus rhythm to aid the doctors for early diagnosis of different cardiovascular abnormalities."

"The presented work will serve as one place assess to available methods and their performance metrics for electrocardiogram analysis that will help the researchers in the future to conduct advanced research in the area of ECG heartbeat classification and further for mental stress estimation derived from ECG."

"The advanced machine learning techniques and customized wearable sensors can be used for real-time analysis and classification of different heartbeats and arrhythmia using electrocardiogram (ECG) signals."

"Researchers have proposed different methods for remote ECG monitoring and classification by combining the data processing capabilities of artificial intelligence and deep learning algorithms together with WSNs and IoT for ECG data acquisition and telemetry [129–134]."

Future Directions

"As a contribution to the present work, the following challenges in future research in the area of ECG signal analysis are identified: Most of the research work used the data sets from the popular MIT-BIH database to evaluate the classification performance which might be significantly different in case of real-time ECG measurements from a set of subjects."

"Researchers have proposed and done the simulation-based validation of machine learning algorithms for fast and accurate classification of ECG beats for remote systems but their application in real-time hospital environments is a challenging area for research because of the complexity and execution time requirements of proposed algorithms."

"For wireless measurement and analysis of real-time electrocardiogram signals, researchers have faced the issues of quality assessment of real-time acquisition, the need of low power consumption nodes, lightweight measurement devices, and simple & fast classification algorithm for long term continuous acquisition of ECG signals required in modern health care systems is a challenging issue."

Forecasting: Bayesian Inference Using Markov Chain Monte Carlo Simulation [135]

This is a machine-generated summary of:

Meenakshisundaram, Swaminathan; Srikanth, Anirudh; Ganesan, Viswanath Kumar; Vijayarangan, Natarajan; Srinivas, Ananda Padmanaban: Forecasting: Bayesian Inference Using Markov Chain Monte Carlo Simulation [135]

Published in: Smart Innovation, Systems and Technologies (2019)
Link to original: https://doi.org/10.1007/978-981-13-5974-3_19

If you want to cite the papers, please refer to the original.

For technical reasons we could not place the page where the original quote is coming from.

Abstract-Summary

"The evolution of modern computing techniques has targeted researchers and technical professionals to delve into an era where forecasting charts future business plans and events."

"Time series plays a vital role in forecasting processes."

"Diverse applications including physics portray time in multiple dimensions."

"Selection and implementation of right forecasting technique require insights into historical data and mapping it with the market expectations which many a time is not aptly forcing for organizations."

"This paper represents various forecasting models and an approach to predict the operational/sales data on a daily basis using combined estimators."

"Markov Chain Monte Carlo (MCMC) has played an extraordinary role in modern engineering applications including economics, physics, statistics and beyond for the past decade."

"Empirical evaluation of various forecasting models resulted in the understanding of the stochastic nature of the processes."

"Out of all the time series models, MCMC yields a satisfactory outcome."

Introduction

"Forecasting a transaction pattern would be highly difficult because it varies from application to application."

"The presence of irregularity can be modelled using various forecasting methods."

"During the late nineteenth century, the economic and civil turbulence was at the peak which led to the foundation of statistical economic forecasting model [136]."

"This model almost ruled all the disciplines as it covered the seasonality and trend, but the trap that led ARIMA model unsuitable is that it accounts only for the stationary time series."

"Kalman filter is one of the most laudable forecasting methods in recent times because it accounts well with inaccuracies and noises that are present in the historical data."

"Various forecasting processes and time series models are discussed in [137]."

"Out of all the models, Bayesian Structural Time Series (BSTS) model performs well as it employs Markov Chain Monte Carlo random sampling method for forecasting."

Literature Review

"Winklhofer and others discussed the importance of forecasting in twenty-first century."

"By Karapanagiotidis [138], typical forecasting methods require information set which forms a linear relationship with other exogenous variables."

"These methods are more suitable for producing point forecasts and it does not provide a computing facility for finding the prediction interval."

"When non-linearity prevailed, ARIMA models provided support to deal linear time series with a better forecast."

"Many forecasters developed ARIMA model into multiple extensions."

"Bayesian forecasting using Markov Chain Monte Carlo (MCMC) has played an extraordinary role in the field of Economics, Physics, Statistics, and beyond for the past decade."

"Before actually starting the forecasting process, it is necessary to analyze the pattern of the original data."

"Data visualization plays a crucial part in forecasting."

"It is advisable to take extra care when implementing forecasting techniques on scarce data."

Design and Underlying Concepts

"Box and Jenkins introduced the ARIMA model in 1970 for the purpose of making a prediction analysis."

"ETS model uses an exponential smoothing procedure to predict its future."

"The ETS model makes use of all the above indicators to make an efficient prediction according to the type of dataset given."

"It is one of the prominent methods that help forecasters make satisfactory predictions when dealing with less amount of data."

"It was mainly developed for fitting complex dynamic models using MCMC sampling."

"The averaging is performed for the predicted values from each model."

"It works by sequentially fitting the model for the poorly predicted observations."

"It predominantly uses Baye's theorem to find uncertainties in forecasting models."

"The detailed description of Bayesian model averaging is discussed in [139]."

Design Consideration

"The first step is to look at the available transaction data for which a month data is chosen for brevity."

"This data enables us to understand the fluctuating cyclic nature of business transaction cycles."

"That the transactions were at the peak during the beginning of the month and then it gradually decreases towards the weekend."

Simulation Procedure

"For testing the simulation results of various forecasting methods, a sample daily volume transaction data is taken from January to December of a year."

"With this data in hand, the data for the past 2 years are calculated by some arbitrary percentage value."

"Let us consider the given data as As discussed earlier, the data for the past 2 years are considered as follows: Since the research is focused on finding the best-suited model in terms of error percentage, an arbitrary value is chosen to find the values of the past years."

"The governing equation for the above-mentioned procedure is given below: where α and β are arbitrary values."

"Reason, the transaction data for the previous years are chosen at random."

"The successful outcome from the exploration stage is fed into forecasting models where all the outcomes are combined into a single optimized estimate."

Results and Evaluation Criteria

"The efficiency of each model is calculated based on the Symmetric Mean Absolute Percentage Error (SMAPE) [140]."

"Forecasting for other models was done."

"It is found that the BSTS results provide enhanced predicted values when compared to other models."

Summary and Future Work

"This paper analyzed the use of various forecasting methods and also depicted the design and outcome of the proposed research."

"This research studied the cyclic patterns of a business transaction and provided a maximum satisfactory forecast of 94.67%."

"The BSTS model yielded a result of 3.86% deviation from the actual observed data."

"The grey models [141] might help the readers to back-forecast the lost data with the present data in hand."

"The mission is aimed to rederive the existing grey model equations in such a way that it yields a maximum acceptable outcome."

References

1. Guo G (2022) Real-time medical image denoising and information hiding model based on deep wavelet multiscale autonomous unmanned analysis. Soft Comput. https://doi.org/10.1007/s00500-022-07322-2
2. Tiwari S, Goel S, Bhardwaj A (2022) EEG signals to digit classification using deep learning-based one-dimensional convolutional neural network. Arab J Sci Eng. https://doi.org/10.1007/s13369-022-07313-3
3. Ghritlahare R, Sahu M, Kumar R (2019) Classification of two-class motor imagery EEG signals using empirical mode decomposition and Hilbert–Huang transformation. In: Computing and network sustainability. Springer, pp 375–386
4. Prpa M, Pasquier P (2019) Brain-computer interfaces in contemporary art: a state of the art and taxonomy. In: Brain art, pp 65–115
5. Gaur P, Pachori RB, Wang H, Prasad G (2018) A multi-class EEG-based BCI classification using multivariate empirical mode decomposition based filtering and riemannian geometry. Expert Syst. Appl. 95:201–211
6. Pratama IWP, Kesiman MWA, Gunadi IGA, Kautsar RJA, Utaminingrum F, Budi AS, Buana W, Meily SO, Fudzee MFBMD, Aminudin A et al (2021) Frequency band and PCA feature comparison for EEG signal classification. Lontar Komput. J. Ilmiah Teknol. Informasi 12(1):1
7. Jolly, B.L.K.; Aggrawal, P.; Nath, S.S.; Gupta, V.; Grover, M.S.; Shah, R.R.. Universal EEG encoder for learning diverse intelligent tasks. In: 2019 IEEE Fifth International Conference on Multimedia Big Data (BigMM), pp. 213–218. IEEE (2019).
8. Dilmi S (2022) Calcium soft sensor based on the combination of support vector regression and 1-D digital filter for water quality monitoring. Arab J Sci Eng. https://doi.org/10.1007/s13369-022-07263-w
9. Pattnaik BS, Pattanayak AS, Udgata SK et al (2021) Machine learning based soft sensor model for BOD estimation using intelligence at edge. Complex Intell. Syst. https://doi.org/10.1007/s40747-020-00259-9
10. Dilmi S, Ladjal M (2021) A novel approach for water quality classification based on the integration of deep learning and feature extraction techniques. Chemom. Intell. Lab. Syst. 214:104329. https://doi.org/10.1016/j.chemolab.2021.104329
11. Zhuiykov S (2012) Solid-state sensors monitoring parameters of water quality for the next generation of wireless sensor networks. Sens. Actuators B Chem. 161(1):1–20
12. Haghiabi AH, Nasrolahi AH, Parsaie A (2018) Water quality prediction using machine learning methods. Water Qual. Res. J. 53(1):3–13
13. Djerioui M, Bouamar M, Ladjal M, Zerguine A (2019) Chlorine soft sensor based on extreme learning machine for water quality monitoring. Arab. J. Sci. Eng. 44:2033–2044. https://doi.org/10.1007/s13369-018-3253-8
14. Qishlaqi A, Kordian S, Parsaie A (2016) Field measurements and neural network modeling of water quality parameters. Appl. Water Sci
15. Mahabub A (2020) Design and implementation of cost-effective IIR filter for EEG signal on FPGA. Aust. J. Electr. Electron. Eng. 17(2):83–91
16. Difference between IIR and FIR filters: a practical design guide. https://www.advsolned.com/difference-between-iir-and-fir-filters-a-practical-design-guide/. Accessed 11 Dec 2020.
17. Ranjan R, Thakur A (2022) Image encryption using discrete orthogonal Stockwell transform with fractional Fourier transform. Multimed Tools Appl. https://doi.org/10.1007/s11042-022-14,240-w
18. Yu C, Li J, Li X, Ren X, Gupta BB (2018) Four-image encryption scheme based on quaternion Fresnel transform, chaos and computer-generated hologram. Multimed Tools Appl 77:4585–4608. https://doi.org/10.1007/s11042-017-4637-6
19. Elkady MM, Abd Elnaby MMA, Hussein AH, Kader HMA (2016) Chaotic based technique for partial image encryption. In: 2016 26th International Conference on Computer Theory and Applications (ICCTA), pp. 20–25. https://doi.org/10.1109/ICCTA40200.2016.9512943

20. Kanso A, Ghebleh M (2012) A novel image encryption algorithm based on a 3D chaotic map. Commun Nonlinear Sci Numer Simul 17:2943–2959
21. Liu Z, Li S, Liu W, Wang Y, Liu S (2013) Image encryption algorithm using fractional Fourier transform and pixel scrambling operation based on double random phase encoding. Opt Lasers Eng 51(1):8–14
22. Ranjan R, Jindal N, Singh AK (2020) Fractional S-transform and its properties: a comprehensive survey. Wirel Pers Commun 113(4):2519–2541. https://doi.org/10.1007/s11277-020-07339-6
23. Wang Y, Orchard J (2008) Symmetric discrete orthonormal Stockwell transform. In: AIP Conference Proceedings 1048:585. https://doi.org/10.1063/1.2990992
24. Wang Y, Orchard J (2009) Fast discrete orthonormal Stockwell transform. SIAM J Sci Comput 31(5):4000–4012
25. Prasad A, Kumar M, Choudhury DR (2008) Color image encoding using fractional Fourier transformation associated with wavelet transformation. Opt Commun 285(6):1005–1009
26. Nithya B, Sankari YB, Manikantan K, Ramachandran S (2015) Discrete orthonormal Stockwell transform based feature extraction for pose invariant face recognition. Procedia Comput Sci 45:290–299
27. Caccamo MT, Giuffrè E, Mezzatesta F, Pizzolo F, Tafaro F, Magazù S (2021) Investigation of a bridge mechanical response by a joint innovative impulsive energizer and a wavelet analysis. Int J Civ Eng. https://doi.org/10.1007/s40999-021-00615-x
28. Sayed MA, Kaloop MR, Kim E, Kim D (2017) Assessment of acceleration responses of a railway bridge using wavelet analysis. KSCE J Civil Eng 21:1844–1853. https://doi.org/10.1007/s12205-016-1762-0
29. Stankovic L, Stankovic S, Dakovic M (2014) From the STFT to the Wigner distribution. IEEE Signal Process Mag 31:163–174. https://doi.org/10.1109/MSP.2014.2301791
30. Cantero D, Ülker-Kaustell M, Karoumi R (2016) Time–frequency analysis of railway bridge response in forced vibration. Mech Syst Signal Process 76–77:518–530. https://doi.org/10.1016/j.ymssp.2016.01.016
31. Caccamo MT, Magazù S (2016) Tagging the oligomer-to-polymer crossover on EG and PEGs by infrared and Raman spectroscopies and by wavelet cross-correlation spectral analysis. Vib Spectr 85:222–227. https://doi.org/10.1016/j.vibspec.2016.04.017
32. El-Borie MA, El-Taher AM, Thabet AA, Bishara AA (2020) The interconnection between the periodicities of solar wind parameters based on the interplanetary magnetic field polarity (1967–2018): a cross wavelet analysis. Solar Phys 295:122. https://doi.org/10.1007/s11207-020-01692-2
33. Gokdag H (2011) Wavelet-based damage detection method for a beam-type structure carrying moving mass. Struct Eng Mech 38:81–97. https://doi.org/10.12989/sem.2011.38.1.081
34. Nguyen KV, Tran HT (2010) Multi-cracks detection of a beam-like structure based on the on-vehicle vibration signal and wavelet analysis. J Sound Vib 329:4455–4465. https://doi.org/10.1016/j.jsv.2010.05.005
35. Khorram A, Bakhtiari-Nejad F, Rezaeian M (2012) Comparison studies between two wavelet based crack detection methods of a beam subjected to a moving load. Int J Eng Sci 51:204–215. https://doi.org/10.1016/j.ijengsci.2011.10.001
36. Bayer FM, Kozakevicius AJ, Cintra RJ (2019) An iterative wavelet threshold for signal denoising. Signal Proc 162:10–20. https://doi.org/10.1016/j.sigpro.2019.04.005
37. Staszewski WJ, Robertson AN (2007) Time-frequency and time-scale analyses for structural health monitoring. Philos Trans R Soc A 365:449–477. https://doi.org/10.1098/rsta.2006.1936
38. Chavez M, Cazelles B (2019) Detecting dynamic spatial correlation patterns with generalized wavelet coherence and non-stationary surrogate data. Sci Rep 9:7389. https://doi.org/10.1038/s41598-019-43571-2
39. Ibrahim M, Jemei S, Wimmer G, Steiner NY, Kokonendji CC, Hissel D (2015) Selection of mother wavelet and decomposition level for energy management in electrical vehicles

including a fuel cell. Int J Hydrogen Energy 40:15823–15833. https://doi.org/10.1016/j.ijhydene.2015.06.055
40. Naresh GJ, Deepak WP (2017) Investigation on selection of optimal mother wavelet in mode shape based damage detection exercise. Proc Eng 181:531–537. https://doi.org/10.1016/j.proeng.2017.02.430
41. Singh A, Maiti S, Tiwari RK (2017) Selection of optimum wavelet in CWT analysis of geophysical downhole data. J Indian Geophys Union 21(2017):153–166
42. Berger P, Wotzka D (2015) Selecting mother wavelet for wavelet analysis of on-load tap changer acoustic emission. Acta Phys Polon A 128:310–314. https://doi.org/10.12693/APhysPolA.128.310
43. Ahuja N, Lertrattanapanich S, Bose NK Properties determining choice of mother wavelet. IEEE Proc Vision Image Signal Process 152:659–664. https://doi.org/10.1049/ip-vis:20045034
44. Büssow R (2007) An algorithm for the continuous Morlet wavelet transform. Mech Syst Signal Process 21:2970–2979. https://doi.org/10.1016/j.ymssp.2007.06.001
45. Caccamo MT, Cannuli A, Magazù S (2018) Wavelet analysis of near-resonant series RLC circuit with time-dependent forcing frequency. Eur J Phys 39:aaae77. https://doi.org/10.1088/1361-6404/aaae77
46. Sanjeet S, Sahoo BD, Parhi KK (2022) Low-energy real FFT architectures and their applications to seizure prediction from EEG. Analog Integr Circuits Signal Process. https://doi.org/10.1007/s10470-022-02094-z
47. Ayinala M et al (2013) An In-place FFT architecture for real-valued signals. IEEE Trans Circuits Systems II Express Briefs 60(10):652–656
48. Yin X et al (2016) Resource-efficient Pipelined architectures for radix-2 real-valued FFT with real datapaths. IEEE Trans Circuits Syst II Express Briefs 63(8):803–807
49. Garrido M et al (2009) A Pipelined FFT architecture for real-valued signals. IEEE Trans Circuits Syst I Regul Pap 56(12):2634–2643
50. Verma N, Shoeb A, Bohorquez J, Dawson J, Guttag J, Chandrakasan AP (2010) A micropower EEG acquisition SoC with integrated feature extraction processor for a chronic seizure detection system. IEEE J Solid-State Circuits 45(4):804–816
51. Wang, H., Shi, W., & Choy, C. (2017). Integrating channel selection and feature selection in a real time epileptic seizure detection system. In: 2017 39th Annual International Conference of the IEEE Engineering in Medicine and Biology Society (EMBC). Seogwipo, pp. 3206–3211.
52. Chandler, D., Bisasky, J., Stanislaus, J.L.V.M., & Mohsenin, T. (2011). Real-time multichannel seizure detection and analysis hardware. In: 2011 IEEE Biomedical Circuits and Systems Conference (BioCAS), San Diego, CA, pp. 41–44.
53. Bisasky, J., Chandler, D., & Mohsenin, T. (2012). A many-core platform implemented for multi-channel seizure detection. In: 2012 IEEE International Symposium on Circuits and Systems (ISCAS). Seoul, pp. 564–567.
54. Altaf, M.A.B., & Yoo, J. (2013). A 1.52 $$\mu $$ μ J/classification patient-specific seizure classification processor using linear SVM. In: 2013 IEEE International Symposium on Circuits and Systems (ISCAS), Beijing, pp. 849–852.
55. Altaf, M.A.B., Tillak, J., Kifle, Y., & Yoo, J. (2013). A 1.83 $$\mu $$ μ J/Classification Nonlinear Support-Vector-Machine-Based Patient-Specific Seizure Classification SoC. In: 2013 IEEE International Solid-State Circuits Conference Digest of Technical Papers, San Francisco, CA, pp. 100–101.
56. Salam MT, Sawan M, Nguyen DK (2011) A novel low-power-implantable epileptic seizure-onset detector. IEEE Trans Biomed Circuits Syst 5(6):568–578
57. Rathi N, Sinha S (2019) Quantitative tissue characterization using discrete wavelet transform of photoacoustic signals: a feasibility study. Signal Image Video Process. https://doi.org/10.1007/s11760-019-01485-6
58. Lopes WN, Junior POC, Aguiar PR, Alexandre FA, Dotto FRL, da Silva PS, Bianchi EC (2021) An efficient short-time Fourier transform algorithm for grinding wheel condition

monitoring through acoustic emission. Int J Adv Manuf Technol. https://doi.org/10.1007/s00170-020-06476-3
59. Griffin JM (2015) Traceability of acoustic emission measurements for micro and macro grinding phenomena???Characteristics and identification through classification of micro mechanics with regression to burn using signal analysis. Int J Adv Manuf Technol 81(9–12):1463–1474
60. Chen X, Öpöz TT (2016) Effect of different parameters on grinding efficiency and its monitoring by acoustic emission. Prod Manuf Res 4(1):190–208
61. Chen X, Mohamed A, Oluwajobi A (2012) Investigation of AE features in grinding. J Phys Conf Ser 364(1):012090. https://doi.org/10.1088/1742-6596/364/1/012090
62. Mohammed A, Folkes J, Chen X (2012) Detection of grinding temperatures using laser irradiation and acoustic emission sensing technique. Mater Manuf Process 27(4):395–400
63. Chen X, Mohamed A, Folkes J (2011) Grinding monitoring through thermal acoustic emission signatures. Adv Mater Res 325:287–293
64. Batako ADL, Goh SY (2014) Workpiece roundness profile in the frequency domain: An application in cylindrical plunge grinding. Int J Adv Manuf Technol 72(1–4):277–288
65. Griffin JM, Chen X (2009) Multiple classification of the acoustic emission signals extracted during burn and chatter anomalies using genetic programming. Int J Adv Manuf Technol 45(11–12):1152–1168
66. Girardin F, Rafieian F, Liu Z, Thomas M, Hazel B (2011) Using angular speed measurement with hall effect sensors to observe grinding operation with flexible robot. Proceedings of Surveillance 6, Compiègne, pp. 1–7.
67. Chen X, Limchimchol T (2006) Monitoring grinding wheel redress-life using support vector machines. Int J Autom Comput 3(1):56–62
68. Lezanski P (2001) An Intelligent system for grinding wheel condition monitoring. J Mater Process Technol 109(3):258–263
69. Moia DFG, Thomazella IH, Aguiar PR, Bianchi EC, Martins CHR, Marchi M (2015) Tool condition monitoring of aluminum oxide grinding wheel in dressing operation using acoustic emission and neural networks. J Braz Soc Mech Sci Eng 63(3):627–640
70. Lopes WN et al (2017) Digital signal processing of acoustic emission signals using power spectral density and counts statistic applied to single-point dressing operation. IET Sci Meas Technol 11(5):631–636. https://doi.org/10.1049/iet-smt.2016.0317
71. Jorgensen EJ, Werner FT, Prvulovic M, Zajić A (2021) Deep learning classification of motherboard components by leveraging EM side-channel signals. J Hardw Syst Secur. https://doi.org/10.1007/s41635-021-00116-2
72. Dogan H, Forte D, Tehranipoor MM (2014) Aging analysis for recycled fpga detection. In 2014 IEEE International Symposium on Defect and Fault Tolerance in VLSI and Nanotechnology Systems (DFT), pp. 171–176.
73. Callan R, Zajić A, Prvulovic M (2013) A practical methodology for measure the side-channel signal available to the attacker for instruction level events. IEEE Micro 14:1–12
74. Tama BA, Vania M, Lee S, Lim S (2022) Recent advances in the application of deep learning for fault diagnosis of rotating machinery using vibration signals. Artif Intell Rev. https://doi.org/10.1007/s10462-022-10,293-3
75. Chai X, Wang Q, Zhao Y, Xin L, Bai O, Yongqiang L (2016) Unsupervised domain adaptation techniques based on auto-encoder for non-stationary EEG-based emotion recognition. Comput. Biol. Med. 79:205–214
76. Shin Y, Lee S, Ahn M, Cho H, Jun SC, Lee H-N (2015) Noise robustness analysis of sparse representation based classification method for non-stationary EEG signal classification. Biomed. Sign. Process. Control 21:8–18
77. Dziedziech K, Jablonski A, Dworakowski Z (2018) A novel method for speed recovery from vibration signal under highly non-stationary conditions. Measurement 128:13–22
78. Iatsenko D, McClintock PV, Stefanovska A (2016) Extraction of instantaneous frequencies from ridges in time-frequency representations of signals. Signal Process 125:290–303

79. Khan NA, Jönsson P, Sandsten M (2017) Performance comparison of time-frequency distributions for estimation of instantaneous frequency of heart rate variability signals. Appl Sci 7(3):221
80. Schmidt S, Heyns PS, De Villiers JP (2018) A tacholess order tracking methodology based on a probabilistic approach to incorporate angular acceleration information into the maxima tracking process. Mech Syst Signal Process 100:630–646
81. Tama BA, Lee SY, Lee S (2020) An overview of deep learning techniques for fault detection using vibration signal. In: INTER-NOISE and NOISE-CON congress and conference proceedings, institute of noise control engineering, pp. 5701–5706.
82. LeCun Y, Bengio Y et al (1995) (1995) Convolutional networks for images, speech, and time series. Handb Brain Theory Neural Netw 3361(10)
83. Nord JH, Nord GD (1995) Mis research: journal status assessment and analysis. Inf Manage 29(1):29–42
84. Liu R, Meng G, Yang B, Sun C, Chen X (2017a) Dislocated time series convolutional neural architecture: an intelligent fault diagnosis approach for electric machine. IEEE Trans Ind Inform 13(3):1310–1320. https://doi.org/10.1109/TII.2016.2645238
85. Chen T, Wang Z, Yang X, Jiang K (2019b) A deep capsule neural network with stochastic delta rule for bearing fault diagnosis on raw vibration signals. Measurement 148:106857. https://doi.org/10.1016/j.measurement.2019.106857
86. Liu H, Zhou J, Zheng Y, Jiang W, Zhang Y (2018) Fault diagnosis of rolling bearings with recurrent neural network-based autoencoders. ISA Trans 77:167–178. https://doi.org/10.1016/j.isatra.2018.04.005
87. Liu J, Zhou K, Yang C, Lu G (2021) Imbalanced fault diagnosis of rotating machinery using autoencoder-based SuperGraph feature learning. Front Mech Eng 16:829–839
88. Zhao X, Jia M, Liu Z (2021) Semisupervised graph convolution deep belief network for fault diagnosis of electormechanical system with limited labeled data. IEEE Trans Ind Inform 17(8):5450–5460
89. Pei X, Zheng X, Wu J (2021) Rotating machinery fault diagnosis through a transformer convolution network subjected to transfer learning. IEEE Trans Instrum Meas 70:1–11
90. Taibi A, Ikhlef N, Touati S (2022) A novel intelligent approach based on WOAGWO-VMD and MPA-LSSVM for diagnosis of bearing faults. Int J Adv Manuf Technol. https://doi.org/10.1007/s00170-022-08852-7
91. Fu W, Shao K, Tan J et al (2020) Fault diagnosis for rolling bearings based on composite multiscale fine-sorted dispersion entropy and SVM with hybrid mutation SCA-HHO algorithm optimization. IEEE Access 8:13086–13104
92. Jin Z, He D, Chen Y et al (2021) Research on fault diagnosis method of train rolling bearing based on variational modal decomposition and bat algorithm-support vector machine. J Phys Conf Ser 1820:012170
93. Cheng H, Zhang Y, Lu W et al (2019) A bearing fault diagnosis method based on VMD-SVD and Fuzzy clustering. Int J Patt Recogn Artif Intell 33:1950018
94. Mohammed H, Rashid T (2020) A novel hybrid GWO with WOA for global numerical optimization and solving pressure vessel design. Neural Comput Appl 32:14701–14718
95. Yan X, Liu Y, Zhang W et al (2020) Research on a novel improved adaptive variational mode decomposition method in rotor fault diagnosis. Appl Sci 10:1696
96. Liang T, Lu H, Sun H (2021) Application of parameter optimized variational mode decomposition method in fault feature extraction of rolling bearing. Entropy 23:520
97. Li H, Fan B, Jia R et al (2020) Research on multi-domain fault diagnosis of gearbox of wind turbine based on adaptive variational mode decomposition and extreme learning machine algorithms. Energies 13:1375
98. He C, Wu T, Gu R et al (2021) Rolling bearing fault diagnosis based on composite multiscale permutation entropy and reverse cognitive fruit fly optimization algorithm—extreme learning machine. Measurement 173:108636

99. Li J, Chen W, Han K et al (2020) Fault diagnosis of rolling bearing based on GA-VMD and improved WOA-LSSVM. IEEE Access 8:166753–166767
100. Liang T, Lu H (2020) A novel method based on multi-island genetic algorithm improved variational mode decomposition and multi-features for fault diagnosis of rolling bearing. Entropy 22:995
101. Babu TN, Ali PSN, Prabha DR, Mohammed VN, Wahab RS, Vijayalakshmi S (2022) Fault diagnosis in bevel gearbox using coiflet wavelet and fault classification based on ANN including DNN. Arab J Sci Eng. https://doi.org/10.1007/s13369-022-06767-9
102. Sharma S, Tiwari SK, Singh S (2019) Diagnosis of gear tooth fault in a bevel gearbox using discrete wavelet transform and autoregressive modeling. Life Cycle Reliab. Saf. Eng. 8(1):21–32. https://doi.org/10.1007/s41872-018-0061-9
103. Huang W, Kong F, Zhao X (2018) Spur bevel gearbox fault diagnosis using wavelet packet transform and rough set theory. J. Intell. Manuf. 29(1):1257–1271. https://doi.org/10.1007/s10845-015-1174-x
104. Li Z, Ma Z, Liu Y, Teng W, Jiang R (2015) Crack fault detection for a gearbox using discrete wavelet transform and an adaptive resonance theory neural network. J. Mech. Eng. 61(1):63–73. https://doi.org/10.5545/sv-jme.2014.1769
105. Wang J, Kong D, Dong S, Wang C (2015) The gearbox fault diagnosis based on wavelet transform. Int. Ind. Inf. Comput. Eng. Confer. 12(1):1973–1976. https://doi.org/10.2991/iiicec-15.2015.430
106. Chen R, Huang X, Yang L, Xu X, Zhang X, Zhang Y (2019) Intelligent fault diagnosis method of planetary gearboxes based on convolution neural network and discrete wavelet transform. Comput. Ind. 106(3):48–59. https://doi.org/10.1016/j.compind.2018.11.003
107. Zhang, H., Zhang, L., Jiang, Y.: Overfitting and underfitting analysis for deep learning based end-to-end communication systems. In: 2019 11th International Conference on Wireless Communications and Signal Processing, pp. 1–6 (2019). https://doi.org/10.1109/WCSP.2019.8927876
108. QMB A, Afzaal M, Qureshi MS, Fayaz M (2021) Machine learning-based EEG signals classification model for epileptic seizure detection. Multimed Tools Appl. https://doi.org/10.1007/s11042-021-10,597-6
109. Bhattacharyya A, Pachori RB (2017) A multivariate approach for patient-specific EEG seizure detection using empirical wavelet transform. IEEE Trans Biomed Eng 64:2003–2015
110. Vidyaratne LS, Iftekharuddin KM (2017) Real-time epileptic seizure detection using EEG. IEEE Trans Neural Syst Rehabil Eng 25:2146–2156
111. Atal DK, Singh M (2019) A hybrid feature extraction and machine learning approaches for epileptic seizure detection. Multidim Syst Sign Process:1–23
112. Gupta V, Pachori RB (2019) Epileptic seizure identification using entropy of FBSE based EEG rhythms. Biomed Signal Process Control 53:101569
113. Mahjoub C, Jeannès RLB, Lajnef T, Kachouri A (2020) Epileptic seizure detection on EEG signals using machine learning techniques and advanced preprocessing methods. Biomed Eng/Biomedizinische Technik 65:33–50
114. Mursalin M, Zhang Y, Chen Y, Chawla NV (2017) Automated epileptic seizure detection using improved correlation-based feature selection with random forest classifier. Neurocomputing 241:204–214
115. Ullah I, Hussain M, Aboalsamh H (2018) An automated system for epilepsy detection using EEG brain signals based on deep learning approach. Expert Syst Appl 107:61–71
116. Park C, Choi G, Kim J, Kim S, Kim TJ, Min K, Jung KY, Chong J (2018) Epileptic seizure detection for multi-channel EEG with deep convolutional neural network. In ICEIC 2018: IEEE International Conference on Electronics, Information, and Communication, pp. 1–5
117. Wang Y, Li Z, Feng L, Zheng C, Zhang W (2017) Automatic detection of epilepsy and seizure using multiclass sparse extreme learning machine classification. Comput Math Methods Med 2017:6849360

118. Saini SK, Gupta R (2021) Artificial intelligence methods for analysis of electrocardiogram signals for cardiac abnormalities: state-of-the-art and future challenges. Artif Intell Rev. https://doi.org/10.1007/s10462-021-09999-7
119. Luz EJS, Schwartz WR, Cámara-Chávez G, Menott D (2016) ECG-based heartbeat classification for arrhythmia detection: a survey. Comput Methods Programs Biomed 127P:144–164. https://doi.org/10.1016/j.cmpb.2015.12.008
120. Francis J (2016) ECG monitoring leads and special leads. Indian Pacing Electrophysiol J 16(3):92–95. https://doi.org/10.1016/j.ipej.2016.07.003
121. Goldberger AL, Goldberger ZD, Shvilkin A (2018) ECG leads. Goldberger's Clin Electrocardiogr. https://doi.org/10.1016/B978-0-323-40169-2.00004-4
122. Shoeibi A (2018) Classification ECG of cardiac signals using LPC features and support vector classification ECG of cardiac signals using LPC features and support vector machine. In: International conference on knowledge -based research in computer engineering & information technology, (March)
123. Swarnkar A, Kumar A, & Khanna P (2014) Performance of wavelet filters for ECG compression based on linear predictive coding using different thresholding functions. In: 2014 international conference on devices, circuits and communications, ICDCCom 2014-proceedings, (4). https://doi.org/10.1109/ICDCCom.2014.7024737
124. Ardhapurkar S, Manthalkar R, & Gajre S (2011) Electrocardiogram compression by linear prediction and wavelet sub-band coding techniques international center of excellence in engineering and management, Aurangabad, India methods. In: 2011 Computing in Cardiology (CinC), pp. 141–144
125. Hassan RF, Shaker SA (2018) ECG signal de-noising and feature extraction using discrete wavelet transform. Int J Eng Trends Technol 63(1):32–39. https://doi.org/10.14445/22315381/ijett-v63p206
126. Mahmoodabadi SZ, Ahmadian A & Abolhasani MD (2005) ECG feature extraction using daubechies wavelets. In: Proceedings of the fifth IASTED international conference on visualization, imaging, and image processing, VIIP 2005, (January 2005), pp. 343–348
127. Pourbabaee B, Roshtkhari MJ, Khorasani K (2018) Deep convolutional neural networks and learning ECG features for screening paroxysmal atrial fibrillation patients. IEEE Trans Syst Man Cybern Syst 48(12):2095–2104. https://doi.org/10.1109/TSMC.2017.2705582
128. Raj S, Ray KC (2017) ECG signal analysis using DCT-based DOST and PSO optimized SVM. IEEE Trans Instrum Meas 66(3):470–478. https://doi.org/10.1109/TIM.2016.2642758
129. Longoni M, Carrera D, Rossi B, Fragneto P, Pessione M, & Boracchi G (2018) A wearable device for online and long-term ECG monitoring. In: IJCAI international joint conference on artificial intelligence, 2018-July, pp. 5838–5840. https://doi.org/10.24963/ijcai.2018/855
130. El Attaoui A, Hazmi M, Jilbab A, Bourouhou A (2019) Wearable wireless sensors network for ECG telemonitoring using neural network for features extraction. Wirel Pers Commun 111(3):1955–1976. https://doi.org/10.1007/s11277-019-06967-x
131. Rachim VP, Chung WY (2016) Wearable noncontact armband for mobile ECG monitoring system. IEEE Trans Biomed Circuits Syst 10(6):1112–1118. https://doi.org/10.1109/TBCAS.2016.2519523
132. Liu C, Zhang X, Zhao L, Liu F, Chen X, Yao Y, Li J (2019) Signal quality assessment and lightweight QRS detection for wearable ECG smartvest system. IEEE Internet Things J 6(2):1363–1374. https://doi.org/10.1109/JIOT.2018.2844090
133. Biagetti G, Crippa P, Falaschetti L, Orcioni S, Turchetti C (2018) Human activity monitoring system based on wearable sEMG and accelerometer wireless sensor nodes. 08 Information and Computing Sciences 0801 Artificial Intelligence and Image Processing 10 Technology 1005 Communications Technologies. Biomed Eng Online 17(S1):1–18. https://doi.org/10.1186/s12938-018-0567-4
134. Dilmaghani RS, Bobarshad H, Ghavami M, Choobkar S, Wolfe C (2011) Wireless sensor networks for monitoring physiological signals of multiple patients. IEEE Trans Biomed Circuits Syst 5(4):347–356. https://doi.org/10.1109/TBCAS.2011.2114661

135. Meenakshisundaram, Swaminathan, Srikanth, Anirudh., Ganesan, Viswanath Kumar., Vijayarangan, Natarajan., Srinivas, Ananda Padmanaban. Forecasting: Bayesian inference using Markov chain Monte Carlo simulation. Research into design for a connected world (2019). doi: https://doi.org/10.1007/978-981-13-5974-3_19
136. Friedman WA (2014) Fortune tellers: the story of America's first economic forecaster. Princeton University Press, Princeton
137. Brockwell PJ, Davis RA (2002) Introduction to time series and Forecasting©. Springer, New York
138. Karapanagiotidis, P.: Literature review of modern time series forecasting methods. 31 Jul 2012
139. Hoeting, J., Madigan, D., Raftery, A., Volinsky, C.: Bayesian model averaging: a practical tutorial. www.stat.colostate.edu/~jah/documents/bma2.ps (1999)
140. Chen C, Twycross J, Garibaldi JM (2017) A new accuracy measure based on bounded relative error for time series forecasting. PLoS ONE 12(3):e0174202. https://doi.org/10.1371/journal.pone.0174202
141. Kayacan K, Ulutas B, Kaynak O (2010) Grey system theory-based models in time series prediction. Expert Syst. Appl. 37(2):1784–1789

Chapter 4
Future Trends

Introduction by the Editor

Any transportation object delays often create exasperation in the transportation industry when not properly mobilized. The increasing development of artificial intelligence models has provoked researchers and scientists to harness modern research problems. This chapter focuses on berth scheduling for vessel arrival prediction, the dynamics of disruptions in air transportation, flight delay predictions, ranking systems based on analytic hierarchical processes, healthcare heart disease prediction and the chemical engineering process of painting pigments. It covers our future needs through CAMS.

In this chapter, we have shown the future trends in supply chains and logistics. The topic of Maritime supply chains and logistics deals with uncertainty issues of deep-sea vessel arrivals at ports. Due to delay on their arrivals, the different stages of planning and supply chain will be affected. To overcome this problem, CAMS applies a dynamic time buffers (DTB) approach.

Analyzing flight delay prediction under concept drift involves understanding how the predictive performance of flight delay models changes over time due to evolving patterns and conditions in the aviation industry. The two major research problems are being solved by CAMS: (i) How do drift handling strategies influence the prediction performance of delays? (ii) Do different scales change the results of drift handling strategies? In the flight delay prediction analysis, different drift handling strategies in aviation under different scales are applied, and their impacts vary according to scale and machine learning models used. Further, we have analyzed US airport delays through disruptive events between 2008 and 2017. Flight delays are a recurring challenge in the air transportation system, often caused by disruptive events like weather disturbances, equipment failures, or congestion. These disruptions create imbalances between system capacity and demand, leading to inefficiencies. Viewing these events through a dynamical systems lens, we analyze cycles of

V. Natarajan (ed.), *Computational Artificial Intelligence and Methods for industries*, https://doi.org/10.1007/978-981-96-5277-8_4

disruption and recovery in airport operations. Utilizing a state-space representation, we capture the severity and spatial impact of delays at US airports from 2008 to 2017. By identifying representative disruption and recovery cycles, we gain insights into common operational patterns. Additionally, we correlate these cycles with specific off-nominal events, such as airport outages, and discern distinct disruption-recovery pathways. Furthermore, we examine temporal trends to understand when and how the system is typically disrupted and subsequently recovers. This research enhances our understanding of the dynamics of airport operations, aiding in the development of strategies to mitigate delays and improve overall system resilience.

Flight delays pose significant financial challenges to airlines, necessitating accurate delay estimates to enhance customer satisfaction and profitability. While numerous research efforts have focused on flight delay modeling and prediction, many solutions fall short in accuracy due to the complexity and volume of data involved. This study proposes a Deep Learning-based approach for aircraft delay anticipation (DL). Deep Learning offers a promising method for addressing complex problems with large datasets, as it can automatically extract key features from data. Moreover, since flight delay data is often noisy, a stack denoising autoencoder technique is integrated into the proposed model. Additionally, the Levenberg-Marquardt algorithm is employed to determine weight and bias fitting values, further enhancing the model's accuracy. Two additional frameworks are developed to evaluate the impact of the stack denoising autoencoder and the LM method on the model architecture. Through experimentation and evaluation, the proposed DL-based approach demonstrates improved performance in predicting flight delays, offering valuable insights for airline operations and management.

In this chapter, we bring out the applications of Internet of Things (IoT) in Airlines. The purpose of this research is to study flight delays from a new angle by utilizing data generated from the emerging Internet of Things (IoT) paradigm. Our primary objective is to improve the understanding of the roots and signs of flight delays as well as discovering related factors. We consider the IoT data generated from distributed sensors that have not been considered in existing works in the analysis of flight delays, and for that purpose, an automatic tool is developed to collect IoT data from various data sources including flight, weather, and air quality index. Based on the heterogeneous data, an algorithm is developed to merge different features from diverse data sources. We adopt predictive modeling to study the factors that contribute to flight delays and to predict the flight delays in the future. The results of our work show a high correlation among the developed features. In particular, the results clearly demonstrate the association between the flight delays and the air quality index factor. The analytic hierarchy process (AHP) methods are used to analyze the consumption of aviation fuel. With the projected air traffic growth, aviation fuel needs will grow by 3% globally per year. Considering this, the aviation industry has set ambitious goals to enhance its fuel efficiency. This study presents an integrated framework for aviation fuel consumption reduction, which will also limit its carbon dioxide emissions. Further, this research aims to categorize influential factors and examine their relative importance for fuel-efficient aviation.

The diagnosis and prognosis of cardiovascular disease are crucial medical tasks to ensure correct classification, which helps cardiologists provide proper treatment to the patient. CAMS applications in the medical niche have increased as they can recognize patterns from data. Using machine learning to classify cardiovascular disease occurrence can help diagnosticians reduce misdiagnosis. This research develops a model that can correctly predict cardiovascular diseases to reduce the fatality caused by cardiovascular diseases. This chapter proposes a method of k-modes clustering with Huang starting that can improve classification accuracy. Models such as random forest (RF), decision tree classifier (DT), multilayer perceptron (MP), and XGBoost (XGB) are used. GridSearchCV was used to hypertune the parameters of the applied model to optimize the result.

CAMS delves into the use of data mining and machine learning techniques to predict heart disease, a leading cause of death globally. The research study focuses on analyzing extensive and complex medical data to aid healthcare professionals in early diagnosis and effective management of heart disease. The research describes various attributes related to heart disease, and the model on the basis of supervised learning algorithms as Naïve Bayes, decision tree, K-nearest neighbor, and random forest algorithm. It uses the existing dataset from the Cleveland database of UCI repository of heart disease patients. The dataset comprises 303 instances and 76 attributes.

In this chapter, we explore the industrial application of machine learning techniques for identifying pure painting pigments using spectral data from both the paint tubes and the artworks of Portuguese artist Amadeo de Souza Cardoso. The research study demonstrates the advantages of using Artificial mixing by Subtractive Methods (ASM) for reference pigments and employs Root Mean Square Error (RMSE) to effectively distinguish mixtures, particularly those containing black and white pigments. Both ASM and RMSE involve simulation and machine learning techniques respectively. In addition, this research brings out the solution to the common problem of highly-imbalanced and limited data in the analysis of historical artwork.

Machine Generated Summaries

Disclaimer: The summaries in this chapter were generated from Springer Nature publications using extractive AI auto-summarization: An extraction-based summarizer aims to identify the most important sentences of a text using an algorithm and uses those original sentences to create the auto-summary (unlike generative AI). As the constituted sentences are machine selected, they may not fully reflect the body of the work, so we strongly advise that the original content is read and cited. The auto generated summaries were curated by the editor to meet Springer Nature publication standards. To cite this content, please refer to the original papers.

Machine generated keywords: flight, delay, flight delay, air, aviation, airport, prediction, airline, delay prediction, weather, delay flight, mining, air traffic, predict, aircraft.

Robust Berth Scheduling Using Machine Learning for Vessel Arrival Time Prediction [1]

This is a machine-generated summary of:

Kolley, Lorenz; Rückert, Nicolas; Kastner, Marvin; Jahn, Carlos; Fischer, Kathrin: Robust berth scheduling using machine learning for vessel arrival time prediction [1]
Published in: Flexible Services and Manufacturing Journal (2022)
Link to original: https://doi.org/10.1007/s10696-022-09462-x

If you want to cite the papers, please refer to the original.
For technical reasons we could not place the page where the original quote is coming from.

Abstract-Summary
"The aim of robust berth scheduling is to derive conflict-free vessel assignments at the quay of a terminal, taking into account uncertainty regarding the actual vessel arrival times which may result from external influences as, e.g., cross wind and sea current."

"The calculation and use of so-called dynamic time buffers (DTBs), which are derived from the different AIS-based forecasts and whose length depends on the estimated forecast reliability, in the berth scheduling model enhance the robustness of the resulting schedules considerably, as is shown in an extensive numerical study."

"The optimization model does not only lead to more robust solutions, but also to less actual waiting times for the vessels and hence to an enhanced service quality, as can be shown by studying the resulting schedules for real vessel data."

"It turns out that the accuracy of the resulting berthing schedules, measured as the deviation of planned and actually realisable schedules, exceeds the accuracy of all forecasts which underlines the usefulness of the DTB approach."

Introduction

"This work is aimed at filling this gap with respect to a specific problem by developing and presenting a new machine-learning-based approach for forecasting vessel arrival times."

"Through the use of real AIS data from the past on which the forecasts are based, but for which the actual vessel arrival times are already known as well, the approach taken in this work allows an evaluation of the forecast quality of the different forecasting methods."

"The approach chosen in this work-the combination of AIS data exploration for forecasting, the subsequent use of an optimization approach which is based on the respective forecasts and the judgement of the quality of the results based on actual vessel arrivals-is, to the authors' knowledge, unique, even if some publications use AIS data for forecasting."

Literature Review

"With respect to the topic of this work, it should be mentioned that De Leon and others [2] as well as Cheimanoff and others [3] and many other authors in the field use randomly generated vessel arrival times in their computational studies, i.e., no real data are used and no arrival time prediction is employed, even in the most recent publications."

"It can be concluded from this literature review that the approaches taken so far in the literature are in many cases based on trajectory or lane prediction from which then an arrival time can be deduced, with Parolas [4] and Grida and Lee [5] being notable exceptions who base their predictions on AIS data."

"Building up on their work, the approach suggested in this paper is the application of different ML methods to AIS data in order to directly derive arrival time predictions from them."

Data Pre-processing

"In terms of data selection, it is recommendable to consider only recent entries from the available historic AIS data, as the overall world fleet's characteristics may change over time, e.g., due to a larger proportion of vessels with higher transport capacities."

"All AIS messages broadcasted with a current position inside the quay area and moreover with the status "5", corresponding to a vessel moored at the quay, are considered to be vessels arriving at the designated container terminal, i.e., a "port event"."

"The first AIS message of each trajectory from inside the quay area ending at the designated container terminal is used to determine the arrival time of that vessel at the port."

"Knowing the actual arrival times of all calling vessels from the past data, the actual remaining travel time can be calculated for each AIS message from the vessels' port approach."

Robust Berth Scheduling

"The robust model with DTB for the BAP with continuous quay (ro-DTB-BAPc) which is presented below is based on the "Model 1" by Liu and others [6] who, as stated above, also aim at a robust berth allocation, taking uncertainty regarding vessel arrival and handling times into account through a scenario-based approach."

"Three aspects of the service quality provided are considered: (1) the vessel waiting time, (2) the deviation from the pre-scheduled berthing position and (3) the non-robust assignments and postponement of vessels, i.e., a high service level."

"The model formulation of the ro-DTB-BAPc can be presented using the notation introduced above:subject to The objective [2] is to minimize all penalty costs resulting from the three aspects of service quality: The potential waiting time, the spatial deviation from the pre-scheduled berthing position and the non-robust assignments or postponement of vessels."

Numerical Experiments

"The number of vessels that are assigned to a robust berthing position by the ro-DTB-BAPc, i.e., that are scheduled according to their DTB, then is used as the desired minimum service level for the benchmark model by Liu and others [6]."

"The ro-DTB-BAPc reaches an average actual waiting time of about 566 min per berthing schedule considering the robustly scheduled vessels, whereas the "Model

1" leads to an actual waiting time of about 902 min on average according to the vessels' actual arrival times."

"Overall the "Model 1" leads to significantly more conflicts per case than the ro-DTB-BAPc with respect to the actual arrival times, providing an average true service level of 65.49% only considering conflicts between the scheduled vessels."

"The actual waiting time is much higher than for the ro-DTB-BAPc, indicating that the vessels often are assigned too late in planning horizon such that the "Model 1" results in a more conservative berthing schedule."

Conclusion and Outlook

"To the best of the authors' knowledge, the approach presented in this work-to develop an AIS data based forecast using ML methods for vessel arrival time prediction at a designated container terminal, which is subsequently used in an optimization model for the robust berth allocation problem, the ro-DTB-BAPc-is unique and fills part of the research gap on data-driven decision making for container terminals as identified by Heilig and others [7]."

"It is to be analysed whether the handling times can be forecasted using ML techniques and the influence on the robustness of the resulting berthing schedules is to be studied."

"It can be evaluated if the accuracy of the vessel arrival time prediction for past data is a good predictor of the future accuracy and hence if more specific weights--i.e., the more successful the method, the larger the weight of the respective forecast-lead to even better results."

Analyzing Flight Delay Prediction Under Concept Drift [8]

This is a machine-generated summary of:

Giusti, Lucas; Carvalho, Leonardo; Gomes, Antonio Tadeu; Coutinho, Rafaelli; Soares, Jorge; Ogasawara, Eduardo: Analyzing flight delay prediction under concept drift [8]

Published in: Evolving Systems (2022)
Link to original: https://doi.org/10.1007/s12530-021-09415-z

If you want to cite the papers, please refer to the original.

For technical reasons we could not place the page where the original quote is coming from.

Abstract-Summary

"This paper investigates the prediction performance of different drift handling strategies in aviation under different scales (models trained from flights related to a single airport or the entire flight system)."

"Two research questions were proposed and answered: (1) how do drift handling strategies influence the prediction performance of delays? (2) Do different scales change the results of drift handling strategies?"

"In our analysis, drift handling strategies are relevant, and their impacts vary according to scale and machine learning models."

Introduction

"Flight data is combined with weather information from departure and arrival locations to help predict flight delays [9, 10]."

"In the former, the goal is to predict the time (usually measured in minutes) that a flight will delay."

"In the latter, the goal is to predict whether the flight is going to delay [11, 12]."

"The literature specialized in flight delay prediction provides many different models that have been developed with good results for both tasks."

"Previous studies indicate that concept drift may impact predictive models [13–15]."

"The amount of data used to train each model (so-called scale) varies across past studies: models trained from flights related to a single airport or the entire flight system."

"Two research questions were proposed and answered: (1) how do drift handling strategies influence the prediction performance of delays? (2) Do different scales change the results of drift handling strategies?"

Background

"Consider a classification problem, such that a set of input variables X is used to predict a class label Y. One of the main challenges when creating machine learning models is handling concept drift."

"There are two main categories of drift detection [16]: (1) data distribution and (2) error rate."

"The detection of drifts can be based solely on data distribution, the error rate of predictions, or both."

"When drift occurs, no action is done, and the trained model might increase its error when predicting newer batches."

"The drawback of this approach is that it might retrain models, even if no drift occurred in the dataset [13, 15]."

"If drift is detected, a new model is built using previous batches."

"A new model (in orange) is used if drift occurs between batches two and three."

Related Work

"Studies that considered streaming approaches were deeply investigated, particularly those that contained drift handling strategies [11, 12, 17–21]."

"Kim et al. [11] used Deep Neural Networks to predict flight delays in the US flights using a BSS of 7 and 9 days before prediction."

"As no specific strategy was used to identify or deal with drifts actively, we considered this approach a passive strategy."

"Pesaranghader et al. [18] tested many drift detection techniques with Naive Bayes and Hoeffding Trees."

"DDM, EDDM, Adaptive Sliding Window (ADWIN), Hoeffding Drift Detection Method (HDDM), and Fast HDDM (FHDDM) active strategies were implemented for comparison reasons."

"The proposed FHDDM uses Hoeffding inequality to compare errors from batches and thus detect drifts."

"And considering the systematic mapping presented here, no study compared active and passive drift handling strategies for delay prediction in aviation."

Methodology

"A second goal is to investigate drift handling strategies under the influence of different scales of training data: System-Based (SB) and Airport-Based (AB)."

"It comprises three steps: (1) preprocessing; (2) drift detection and handling; (3) model training and evaluation."

"The active strategy compared the current training batch sequence with the previous one to detect a drift."

"The function that implements all drift features and returns whether a model should be trained and the corresponding training data is described in Algorithm 2."

"The two most recent batches considering the t parameter are created from the full dataset D. Then, the drift detection (dd) and drift handling (dh) parameters are used to check the presence of drifts between the two most recent batches."

"The function cloakModel receives the parameters from the checkDrift function as well as the airport (airport), machine learning model (mlm), time (t), BSS (b), drift detection method (dd) and drift handling strategy (dh)."

Experimental Evaluation

"A top-k analysis of the best combinations of parameters (drift handling strategy, drift detection method, classifier, and BSS) was also executed."

"For Active strategies, all drift detection methods showed similar results."

"These results may indicate that more retrains may be better for f1 performance for datasets with a high number of drifts."

"To analyze the prediction performance of all parameters used in Algorithm 1, we conducted multiple linear regressions for each airport with the Drift Strategy-Detection combination, classifier, and BSS to estimate F1."

"These results may indicate that increasing the size may not be relevant to the f1 score, and the 1-year size may be the best choice considering the values we tested."

"Considering that BSS higher than one does not significantly increase the prediction performance, an essential gain in execution time is possible."

Conclusions

"We analyzed different types of drift handling strategies in aviation."

"Two research questions were answered to achieve the main objective of this study: (1) how do drift handling strategies influence the prediction performance of delays? (2) Do different scales change the results of drift handling strategies?"

"Choosing machine learning models may have a higher impact on f1 than drift handling strategies."

"Future studies may consider testing error-based drift handling strategies, investigating ensemble drift detection methods, implementing more robust classifiers (CNN, LSTM, and others), check drift handling strategies under different thresholds for flight delays, and evaluate other metrics such as the ROC curve."

Dynamics of Disruption and Recovery in Air Transportation Networks [22]

This is a machine-generated summary of:

Li, Max Z.; Gopalakrishnan, Karthik; Balakrishnan, Hamsa; Shin, Sang Hyun; Jalan, Darsh; Nandi, Aritro; Marla, Lavanya: Dynamics of disruption and recovery in air transportation networks [22].

Published in: CEAS Aeronautical Journal (2021)
Link to original: https://doi.org/10.1007/s13272-021-00521-x

If you want to cite the papers, please refer to the original.
For technical reasons we could not place the page where the original quote is coming from.

Abstract-Summary
"These cycles of disruptions and subsequent recoveries can be viewed from a dynamical systems perspective: exogenous inputs (convective weather, airspace restrictions, etc) disrupt the system, inducing delays and inefficiencies from which the system eventually recovers."

"We study these disruption and recovery cycles through a state-space representation that captures the severity and spatial impact of airport delays."

"Using US airport delay data from 2008 to 2017, we first identify representative disruption and recovery cycles."

"These representative cycles provide insights into the common operational patterns of disruptions and recoveries in the system."

Introduction

"Disruptions and subsequent recoveries in the air transportation system vary in their geographical extent (number of airports affected), intensity (severity of resultant delays and cancellations), and duration (ranging from hours to days)."

"Our work focuses on formalizing and analyzing disruptions and recoveries in the air transportation system by leveraging techniques that evaluate aviation disruptions from the perspective of signal processing in networked systems [23]."

"Prior work has defined aviation system recovery in terms of flight delays, displaced or delayed crews, and disrupted or delayed passengers [24, 25]."

"One of our key technical contributions is the choice of incorporating spectral graph-theoretic and temporal features to describe disruptions and recoveries in airport networks."

"We uncover and interpret two interesting observations related to: (1) the behavior of disruption and recovery during off-nominal events (e.g., airport outages), and (2) the temporal trends in disruption–recovery trajectories."

Data and Methodology

"A symbiotic maneuver is one in which both TD and TV are increasing, or that both are decreasing."

"A trade-off maneuver is one in which the TV and TD change in opposite directions."

"A trade-off maneuver could indicate that although the system delay is decreasing, its spatial variability is increasing."

"Among the remaining maneuvers, symbiotic maneuvers with increasing TV and TD are defined as a disruption segment, and symbiotic maneuvers with decreasing TV and TD are defined as a recovery segment."

"The four possible trade-off maneuvers are defined as follows: i. Disruption-in-TV segment: Increasing TV and decreasing TD following a disruption."

"Disruption-in-TD segment: Increasing TD and decreasing TV following a disruption."

"Recovery-in-TD segment: Increasing TV and decreasing TD following a recovery."

"Recovery-in-TV segment: Increasing TD and decreasing TV following a recovery."

Representative DRTs

"To the other short representative DRT (Short_Rec), the average area for Short_Dis is positive, indicating that the airport delays were spatially distributed in a more unexpected manner during disruption than recovery."

"Similar to Short_Dis, these DRTs represent transient off-nominal conditions, with an average length of 4 h. The average TD and TV intensity values for Short_Rec are smaller than those of Short_Dis, indicating that the system state does not typically attain the maximum TD and TV values."

"Operational day-long DRTs with spatially perturbed disruption segments (OpsDay_Dis): With average DRT lengths of approximately 15 h, these disruptions and subsequent recoveries account for a major portion of an operational day."

"Operational day-long DRTs with spatially perturbed recovery segments (OpsDay_Rec): The temporal persistence of these DRTs is similar to OpsDay_Dis, with an average length of 18 h. As we have noted in OpsDay_Dis, the spatial delay distribution for OpsDay_Rec is more unexpected during recovery segments."

Evaluating Off-Nominal Events and Temporal Trends

"Long-lasting DRTs (i.e., OpsDay_Rec, OpsDay_Dis, and MultiDay) are present in over 70% of nor'easter- and thunderstorm-type days, but only around 37% of days with an airline- or airport-specific outage."

"37% of DRTs during nor'easters are of type OpsDay_Rec, and 44% of DRTs during thunderstorms are MultiDay."

"Examining the average representative DRT lengths, the time it takes the system to be disrupted and recover from nor'easter-type days tends to be shorter than for thunderstorm-type days, which are dominated by MultiDay DRTs."

"63% of DRTs are Med-length or shorter on outage-type days, indicating that the disruption and recovery of the system during these events are short-lived."

"Most DRTs (84%) during hurricane-type days are short-term disruptions and recoveries."

"MultiDay DRTs in April and July 2017 are predominantly thunderstorm-type off-nominal days."

Concluding Remarks

"This paper defined disruptions and subsequent recoveries using the delay magnitude (TD) and the spatial distribution (TV) of the delays, in conjunction with a low-dimensional state-space trajectory representation."

"We then focused on the problem of finding representative DRTs."

"The next steps will be to examine each of the individual DRTs at a more microscopic level, with a focus on time periods involving trade-off maneuvers."

"These eigenvectors are closely related to the TV, and demarcate the airport delays that contribute to unexpected spatial delay distributions [23, 26]."

"Another direction of future research is to leverage the representative DRTs as features for predicting future system behavior, both at the system-wide and airline-specific levels."

Optimization of Air Traffic Management Efficiency Based on Deep Learning Enriched by the Long Short-Term Memory (LSTM) and Extreme Learning Machine (ELM) [27]

This is a machine-generated summary of:

Yousefzadeh Aghdam, Mahdi; Kamel Tabbakh, Seyed Reza; Mahdavi Chabok, Seyed Javad; Kheyrabadi, Maryam: Optimization of air traffic management efficiency based on deep learning enriched by the long short-term memory (LSTM) and extreme learning machine (ELM) [27].
Published in: Journal of Big Data (2021)
Link to original: https://doi.org/10.1186/s40537-021-00438-6

If you want to cite the papers, please refer to the original.

For technical reasons we could not place the page where the original quote is coming from.

Abstract-Summary

"Various methods (e.g., statistical and fuzzy techniques) and data mining algorithms (e.g., neural network) have been used to solve the issues of air traffic management (ATM) and delay the minimization problems."

"To increase the air traffic management accuracy and legitimacy we used the bidirectional long short-term memory (Bi-LSTMs) and extreme learning machines (ELM) to design the structure of a deep learning network method."

"The proposed method has had a % increase in air traffic management in comparison to other papers."

"It can be said that the proposed method has a much higher air traffic management capacity in comparison to the previous methods."

Introduction

"The air traffic control methods for the prevention of flight delay and interference is a significant issue in the operating field of ATM [28]."

"The machine learning algorithm is a conventional method which used to resolve the issues of air traffic, ASL, delay forecast, and minimization."

"This novel technique could solve problems with a high accuracy owing to its ability to accept the large data of the problem and neural network integration, as well as learning techniques and structural dynamism, in the hidden layers formation."

"Aviation and ATM issues are no exception, and most of the recent studies regarding flight delay forecast and flight control traffic have benefited from this technique [28–32]."

"Multilayer neural networks or deep neural networks are included in machine learning subject and a set of algorithms that attempt to model high-level abstract concepts based on learning at various levels and layers, thereby enabling deep learning to process large volumes of data in complicated categories [28, 30, 31, 33–35]."

A Review of Previous Methods

"In [36], the researchers proposed a deep learning-based model to predict the hard landings of aircrafts based on the quick access record (QAR)."

"In [37], the main objective was to predict flight routes using a deep neural network in a capacity management and air traffic operational system."

"In [38], an end-to-end deep learning-based approach was also presented to increase accuracy in the air traffic flow using the CNN and RNN algorithms, as well as a convolutional LSTM module, which was proposed to construct a trainable model to predict the air traffic flow."

"In [39], a multi-step deep sequence learning model (Bi-LSTM + Seq2Seq) was proposed to predict airport delay based on the spatial and temporal relations of the other airports within the network."

"At the next stage, the Bi-LSTM + Seq2Seq model was proposed and trained based on the time-series data on the current status of the network and delay in the interactions between the airports."

Proposed Method

"Column-row-based Soft max functions were used to generate accuracy weights for the hidden modes of the problem and response separately by the following equation: In order to obtain the accuracy vector of the question with respect to each word of the response, we combined the explanatory weights and approximation matrix to calculate the new CQ and CA field vectors."

"While the input weight and bias of the hidden layer of the ELM network are generated randomly, the weights obtained from the neural network encompassing BiLSTM was applied at this stage of the present study, and only the output weight had to be determined, which could limit the manual adjustment of the parameters of each layer in the BP neural network and improve the predictive accuracy."

"X_1, x_2,...,x_n were the input of the educational data, and w_{ij} and β_{jk} were the input weight in the neural network and indicative of the output weight vector between the hidden layer and output node, respectively."

Analysis and Evaluation

"Delays due to flight operation conditions at the airport; 3."

"Delays due to the flight atmosphere Since the type of delay in the present study included the numbers one (delays due to flight planning and scheduling), two (delays due to flight operation conditions at the airport), and six (delays due to the flight conditions of the national flight network), the amount of delay time slot was considered to be less than 15 min and 15–30 min based on the mentioned findings."

"The accuracy of the 30-min delays was higher at the ATL Airport."

"In delays of 30 min and more, the recorded information is more accurate since the order of flight arrival and departure numbers changes according to the order intended for the flight with the airport control mechanism and it is necessary to send a flight delay message or a flight plan update."

Conclusion

"The improved accuracy in ATM management problems is proposed in this paper."

"The dual-sided LSTM algorithm is used to improve the 15 and 30-min delays' accuracy."

"The results show a higher accuracy improve the rate in comparison to other paper and also show that the RMSE parameter in 30-min delays has a lower percentage in ORD, PHX, and JFK airports in comparison to the other airports."

"In further studies, to increase the ATM accuracy, other LSTM models like Casc-LSTM and Ens2-LSTM can be used alongside the ELM algorithm."

Flight Delay Prediction Based on Deep Learning and Levenberg-Marquart Algorithm [40]

This is a machine-generated summary of:

Yazdi, Maryam Farshchian; Kamel, Seyed Reza; Chabok, Seyyed Javad Mahdavi; Kheirabadi, Maryam: Flight delay prediction based on deep learning and Levenberg-Marquart algorithm [40]

Published in: Journal of Big Data (2020)
Link to original: https://doi.org/10.1186/s40537-020-00380-z

If you want to cite the papers, please refer to the original.

For technical reasons we could not place the page where the original quote is coming from.

Abstract-Summary

"An accurate estimation of flight delay is critical for airlines because the results can be applied to increase customer satisfaction and incomes of airline agencies."

"There have been many researches on modeling and predicting flight delays, where most of them have been trying to predict the delay through extracting important characteristics and most related features."

"Most of the proposed methods are not accurate enough because of massive volume data, dependencies and extreme number of parameters."

"This paper proposes a model for predicting flight delay based on Deep Learning (DL)."

"Due to the fact that most of flight delay data are noisy, a technique based on stack denoising autoencoder is designed and added to the proposed model."

"In order to study effect of stack denoising autoencoder and LM algorithm on the model structure, two other structures are also designed."

"To investigate the three models, we apply the proposed model on U.S flight dataset that it is imbalanced dataset."

"Accuracy of the proposed prediction model analyzed and compared to previous prediction method."

"results of three models on both imbalanced and balanced datasets shows that precision, accuracy, sensitivity, recall and F-measure of SDA-LM model with imbalanced and balanced dataset is improvement than SAE-LM and SDA models."

"The results also show that accuracy of the proposed model in forecasting flight delay on imbalanced and balanced dataset respectively has greater than previous model called RNN."

Introduction

"According to abundant and diversity of reasons for flight delays, We are faced with a massive amount of data which is not possible to be processed through previous methods of data [41] analysis like classification [42], or the decision tree [43] and machine learning based methods [41, 42, 44–46] to process this volume of data are not proper, because characteristics of older intelligent system has been designed by human and usually were personalized, also people rarely perceive some features and usually neglect these matters."

"Model, the destination delay is highly dependent to arrival flights and the effective factors include; day, time, airport capacity and some factors are related to passenger loads."

“This model employs factors of airport type, airplane type, date, time, flight path, flight frequency for network training and non-linear and linear for data analysis.”

“One of the recently presented works in solving problem employs the recurrent Deep Neural Network and its results has a high accuracy in flight delay prediction [47].”

“In this paper by developing a deep learning-based model, the accuracy of flight delay predictions can be increased.”

Literature Review

“One of the best studies [48] in the area of machine learning method has been presented by a model which applicate machine learning techniques to investigate delay in arrival flights.”

“the researcher has presented a model based on support vector regressor (SVR) algorithm to predict flight delay in U.S.A airports.”

“A group of researchers [49] have designed 5 models to predict flight delay based on machine learning models such as Logistic Regression, Decision Tree Regression, Bayesian Ridge, Regression and Gradient Boosting Regression.”

“Has been evaluated the prediction results using Root Mean Square Error (RMSE) to measure the performance of flight delay time prediction using different models and different combinations of data sources.”

“Some Research [50] combined a deep belief network with a support vector machine to create a prediction model (DBN–SVR), in which DBN extracted the main factors with tangible impacts on flight delays, reduced the dimension of inputs, and eliminated redundant information.”

Methodology

“After adding a denoising autoencoder to the network, previous hidden layer information is transferred as an input to this layer and non-linear transmission among consecutive hidden layers cause learning of structure and next, the resulted network could foresee flight delay.”

“After determining the output, with its help, network error could be computed and then return the value along with Cost Function chart back to the network to update network’s Weights.”

“In back-propagation, network’s movement is backwards, errors and charts return to the hidden layer so that Weights are updated based on them.”

“This algorithm has two main phases [51]; Computing the Jacobian Matrix that is the most complicated part of this algorithm, and calculate the Hessian matrix and updating Weights which Network’s error is computed in this phase.”

"After determine optimized values for weight and bias, we expect network's foresight improve and the proposed model get as close as possible to reality."

Results and Discussions

"In order to evaluate the model, the number of denoising autoencoders and neurons must be determined based on the values for precision, accuracy and time consuming."

"In order to do this, at first, the model is trained using one stack and 64 neurons, and the precision and accuracy values are calculated."

"By adding another denoising autoencoder, the values for precision and accuracy are increased; therefore, another stack was added to the model's structure."

"When the number of neurons increases from 16 to 32 and from 32 to 64, the values for precision and accuracy increased for both datasets; however, by increasing the number of neurons from 64 to 128, the precision and accuracy of the model increased in the training set while they decreased in the testing set."

"Effect of stack denoising autoencoder on noisy data is positive and increase in precision and accuracy in the structure."

Conclusion

"Comparing the three models for two of imbalanced [52] and balanced datasets shows that accuracy of SDA-LM model with imbalanced dataset respectively is greater by 8.2% and 11.3% Than SAE-LM and SDA models."

"Using stack denoising autoencoder and LM algorithm in optimizing the results, and also balancing the dataset, has positive effect on delay forecasting and leads to increment in accuracy and precision of SDA-LM model with imbalanced dataset is greater by 6.1% and 5.4% than SAE-LM and SDA models."

"Whereas, the accuracy of the SDA-LM model with balanced dataset is greater by 10% than SAE-LM and SDA models and the amount of precision is the same for all three models with balance dataset."

"Using our experimental results, we show that accuracy of the model on imbalanced dataset is 92.1% and for balanced dataset is 96.2%, which is respectively greater by 4.1% and 8.2% respectively."

Flight Delay Prediction Using Deep Convolutional Neural Network Based on Fusion of Meteorological Data [53]

This is a machine-generated summary of:

Qu, Jingyi; Zhao, Ting; Ye, Meng; Li, Jiayi; Liu, Chao: Flight Delay Prediction Using Deep Convolutional Neural Network Based on Fusion of Meteorological Data [53].
Published in: Neural Processing Letters (2020)
Link to original: https://doi.org/10.1007/s11063-020-10318-4

If you want to cite the papers, please refer to the original.
For technical reasons we could not place the page where the original quote is coming from.

Abstract-Summary
"To make full use of the characteristics of flight data and meteorological data, two flight delay prediction models using deep convolution neural network based on fusion of meteorological data are proposed in this paper."

"Flight data and meteorological data are fused in the model."

"Both DCNN and SE-DenseNet models are used to extract feature automatically based on the fused flight data set."

"For proposed DCNN model, both straight channel and convolution channel are designed to guarantee the lossless transmission of the feature matrix and enhance the patency of the deep network."

"For proposed SE-DenseNet model, a SE module is added after the convolution layer of each DenseNet block, which can not only enhance the transmission of deep information but also achieve feature recalibration in the feature extraction process."

"The two deep convolutional neural networks proposed in this paper, DCNN and SE-DenseNet, can both effectively improve the prediction accuracies, reaching to 92.1% and 93.19%, respectively."

Introduction

"Cao and others [54] proposed a flight delay prediction method based on Bayesian network."

"Khanmohammadi and others [55] adopted binary coding of flight data, and then input it into artificial neural network for delay prediction."

"The first one is that the traditional flight delay prediction methods are only trained by small data sets, so the prediction accuracy is difficult to improve."

"In order to improve the prediction accuracy, two flight delay prediction models based on deep convolution neural network are proposed."

"Traditional CNN models cannot solve the problems effectively even if the networks are deepened, so a flight delay prediction model based on DCNN (Dual-channel Convolutional Neural Networks) is proposed in this paper, referring to the ResNet network structure."

"The flight delay prediction methods based on deep convolution neural networks can greatly improve the prediction accuracy."

Dual-Channel Convolutional Neural Network

"As the traditional CNN (Convolutional Neural Network) model cannot effectively solve this problem even when the network is deepened, a flight delay prediction model based on DCNN (Dual Channel Convolutional Neural Network) is proposed."

"Convolution attenuation factor was used to control the proportion of feature matrix from different depth and guarantee the steady of the model."

"In DCNN, each block contains two convolution modules and each convolution module accumulates once with the straight channel, which can fuse the features extracted from different network depths."

"In order to ensure the smooth transmission of information in the network, the convolution attenuation factors are introduced to restrict the sparsity of the output characteristic matrix of the convolution channel."

"When training convolution neural network, in forward propagation, the output characteristic matrix of each layer is calculated one by one and the loss function is calculated at last."

SE-DenseNet

"The proposed SE-DenseNet combines the advantages of DenseNet and SENet, which can effectively use multi-dimensional feature information, strengthen the transmission of deep information and enhance the learning and expression ability of deep network through the strategy of "feature recalibration"."

"It can also slow down the attenuation of error items in each hidden layer, ensure the stability of gradient weight information and avoid the disappearance of gradient through the reverse conduction mechanism of the network itself, so as to further improve the network performance."

"The deeper SE-DenseNet is stacked through the above feature transformation layer and cascaded among several structural blocks."

"The training of SE-DenseNet model mainly transmits the error information of training samples back to the hidden layer by Back Propagation (BP) algorithm."

"Based on BP algorithm, the gradient values of each hidden layer are deduced by taking SE-DenseNet's first network block as an example."

Model Optimization

"BN (Batch normalization) technology can speeds up network training [56]."

"The basic idea is to normalize the output data of each layer when each batch of samples is used to train the network."

"Padding can enhance the feature extraction of edge data and control the dimension of flight delay data, so that the deep network of feature extraction can be trained."

"If padding technology is not adopted, the dimension of feature extraction will rapidly decrease and the deep network can not be trained."

Flight Delay Prediction

"Meteorological data M is the weather condition of a flight taking off and landing airport, which mainly include 12 attributes, such as wind direction, wind speed, visibility, temperature, air pressure, the airport where the observation station is located, observation time and so on."

"In spatial dimension, the destination airport in the flight data and the airport in the meteorological data are used as correlation parameters."

"The processed data is transformed into a feature matrix suitable for input into a CNN."

"The one-dimensional flight feature data is transformed in size to make it easy to input into the neural network."

"The flight feature data used in the experiment was 3,955,056, the feature dimension was 81, and the total dimension of the feature number in the input network after matrix transformation was $3{,}955{,}056 \times 9 \times 9$."

The Discussion About Simulation Results

"In order to further verify the stability and training depth of the DCNN and SE-DenseNet model, 88, 100 and 124 layers of network are trained."

"Data set, when the network layer is 124, the accuracy of DCNN and SE-DenseNet can achieve about 92.26% and 93.19%, respectively."

"If the gradient value of the first hidden layer of the network keeps stable with the number of iterations, the model can guarantee the effective transmission of information in each hidden layer of the network."

"The flight delay prediction model based on DCNN can obtain higher accuracy because of the lossless transmission of forward information and the stability of reverse gradient."

"If it is set too small, it will reduce the return gradient value, slow the convergence of the network, and increase the training time of the model."

"Compared with other methods, DCNN and SE-DenseNet models based on deep learning have higher prediction accuracy."

Conclusion

"Two flight delay prediction using deep convolutional neural network based on fusion of meteorological data are proposed."

"When the network deepens, the performance of the model using CNN becomes worse, while the model using DCNN remains stable, which can get higher accuracy. (2) The deep SE-DenseNet can not only alleviate the problem of the disappearance of the deep network gradient, but also realize the adaptive weight calibration of the feature channel, and ultimately improve the overall performance of the network. (3) Compared with the traditional algorithm, the methods based on deep convolution neural network can fully excavate the hidden value of flight and weather factors without feature dimensionality reduction, and achieve higher prediction accuracy. (4) After fusing meteorological data, the performance of DCNN and SE-DenseNet has been improved, which proves that meteorological data is very important for flight delay prediction."

A System for Effectively Predicting Flight Delays Based on IoT Data [57]

This is a machine-generated summary of:

Aljubairy, Abdulwahab; Zhang, Wei Emma; Shemshadi, Ali; Mahmood, Adnan; Sheng, Quan Z.: A system for effectively predicting flight delays based on IoT data [57]

Published in: Computing (2020)
Link to original: https://doi.org/10.1007/s00607-020-00794-w

If you want to cite the papers, please refer to the original.

For technical reasons we could not place the page where the original quote is coming from.

Abstract-Summary

"Due to the highly dynamic environments of the aviation industry, relying only on historical datasets of flight delays may not be sufficient and applicable to forecast the future of flights."

"The purpose of this research is to study the flight delays from a new angle by utilising data generated from the emerging Internet of Things (IoT) paradigm."

"Our primary goal is to improve the understanding of the roots and signs of flight delays as well as discovering related factors."

"We present a framework that aims at improving the flight delay problem."

"We consider the IoT data generated from distributed sensors that have not been considered in existing works in the analysis of flight delays, and for that purpose, an automatic tool is developed to collect IoT data from various data sources including flight, weather, and air quality index."

"We adopt predictive modelling to study the factors that contribute to flight delays and to predict the flight delays in the future."

"The results clearly demonstrate the association between the flight delays and the air quality index factor."

Introduction

"To the flight delay analysis in previous studies that rely mainly on historical data, this study is based on a new type of data obtained from the IoT data sources including flights, weather, and air quality sensors."

"Tackle a number of technical challenges to enable more effective flight delay analysis based on the IoT data."

"This study is one of the first to examine the connection between contextual IoT data and the flight delays [58]."

"The results demonstrate the benefits of utilizing IoT data for tackling the flight delay problem."

"We conduct correlation analysis among features extracted from various data sources to identify the most significant factors related the flight delay problem."

"Results suggest that relying on real-time IoT data provides better understanding of flight delays."

Related Work

"In [59], the authors studied major factors that contribute to flight delays, and also developed a model so as to predict flight delays using the historical records of the Denver International Airport, Colorado, USA."

"The historical data of flights, weather, information on the airlines, and propagation of delays were taken into consideration for training the respective model."

"In [60], a state-of-the-art predictive modeling approach has been employed on airlines' data sets to evaluate the feasibility of machine learning techniques for improvement in the service quality (flight delay prediction) and cost reduction (fuel consumption prediction)."

"A novel prediction system for predicting the estimated time of arrival of commercial flights has been presented in [61], wherein, historical trajectories (along with their 3D grid points) have been employed in order to collect key features, including but not limited to, weather parameters, air traffic, and airport data along the potential flight's path."

Our Approach

"We use flight, weather, and air quality index as the application terms to look for the data sources that contains maps."

"Additional IoT data sources could be identified and exploited if they are relevant to the flight delay problems."

"To that, schedule and flight status data such as the flight number, the origin, the destination, the scheduled and actual departure time, the scheduled and actual arrival time, the aircraft type, and many other flight details are retrieved from airlines and airports and aggregated together with the previous data sources."

"We develop a crawler engine for collecting real-time data from IoT data sources."

"After collecting the data from the data sources, the crawler integrates this data into one dataset."

"For the integration algorithm, in line 1, we provide the data sources to be merged together."

Experimental Studies

"We have extensively studied several countries as case studies and report the experimental results using the data from China due to its quickly-growing aviation industry and poor records of flight delays."

"In the case of retrieving data from WeatherUnderground and Real-time Air Quality Index, we do not know the locations and the number of the IoT sensors around each airport."

"This study focuses on the evaluation of predicting the flight delays using IoT data."

"The results indicate that using features from all the three datasets (flight, weather, and aqi) tends to provide better prediction."

"We utilize real-time data provided by distributed and publicly available IoT sensors including flight, weather, and air quality index."

"We investigate features from the three data sources and study their impact on the delay at departure (DAD)."

"We create multiple logistic regression models to classify flights based on a set of features composed from the three IoT data sources."

Conclusion

"We present a new approach for flight delay prediction by exploiting the Internet of Things (IoT) data."

"We implement an algorithm to effectively discover and merge different features from diverse IoT data sources."

"The extensive experimental results demonstrate the benefits of utilizing IoT data for tackling the flight delay problem."

"For the future work, we plan to include more IoT data sources related to the airports to improve the effectiveness of the approach."

Semantic-Aware Aircraft Trajectory Prediction Using Flight Plans [62]

This is a machine-generated summary of:

Georgiou, Harris; Pelekis, Nikos; Sideridis, Stylianos; Scarlatti, David; Theodoridis, Yannis: Semantic-aware aircraft trajectory prediction using flight plans [62]
Published in: International Journal of Data Science and Analytics (2019)
Link to original: https://doi.org/10.1007/s41060-019-00182-4

If you want to cite the papers, please refer to the original.
For technical reasons we could not place the page where the original quote is coming from.

Abstract-Summary
"Aircraft trajectory prediction (TP) is a challenging and inherently data-driven time-series modeling problem."

"Flight plans, localized weather and aircraft properties are introduced as trajectory annotations that enable modeling in a space higher than the typical 4-D spatio-temporal."

"A multi-stage hybrid approach is employed for a new variation of the core TP task, the so-called Future Semantic Trajectory Prediction, including clustering the enriched trajectory data using a semantic-aware similarity function as distance metric."

"A separate predictive model is trained for each cluster, using a nonuniform graph-based grid that is formed by the waypoints of each flight plan."

"Flight plans constitute a constrained-based training of each predictive model, one for each waypoint, independently."

Introduction

"In review, the major research challenges addressed in this work are: (a) design high-performance, modular, parallelizable trajectory prediction algorithms for the aviation domain in the context of big data; (b) exploit enrichments of surveillance data, e.g., weather, aircraft type; (c) improve the accuracy of the current state of the art in terms of full-flight prediction error by using the latest submitted flight plans as input."

"The merits and contributions of our work are summarized as follows: We define the FSTP problem and propose a multi-stage 'pipelined' processing, with clustering in the first-stage dealing with dimensionality reduction from the N-dim enriched trajectory space to a set of 'similar' (compact) clusters upon which the predictors are trained; We devise a novel predictive modeling representation for semantic trajectories, exploiting flight plans as input; We introduce an inherently parallelizable design with separate lightweight predictors per reference waypoint at the subsequent stages, working with on a nonuniform graph-based grid of much lower complexity compared to the current state-of-the-art practices; We provide an extensive experimental study with real aviation surveillance data (IFS radar tracks, flight plans, weather, etc), assessing the overall prediction accuracy of the proposed approach in contrast to current state-of-the-art alternatives, illustrating at least a sixfold improvement."

Related Work

"In the context of trajectory prediction, the flight route and all the associated information (weather, aircraft properties, etc) are usually encoded into discrete values that constitute the HMM states; then, the trajectory itself is treated as an evolution of transitions between these states, using the raw trajectory data of a large set of flights for training, plus spatio-temporal constraints (locality) to reduce the dimensionality of the problem."

"Ayhan and Samet [63] introduce a novel stochastic approach to aircraft trajectory prediction problem, which exploits aircraft trajectories modeled in space and time by using a spatio-temporal data grid."

"A stochastic approach such as the HMM can address the TP problem by taking environmental uncertainties into account and training a model, using historical trajectory data along with weather observations."

"Despite the plethora of various TP approaches in the aviation domain, most of them are either based on discovered mobility patterns from historical data or 'blind' predictive modeling along entire flight paths in the sense that no constraint is imposed to the search space."

Problem Formulation

"A flight plan announced by an airline is a low-resolution raw trajectory that consists of the waypoints and times that the aircraft is constrained to pass through them."

"In order for a trajectory R to be comparable with a flight plan F, the latter should be enriched/annotated in a similar way, to encapsulate the same information as the semantic trajectories (e.g., weather information is attached to the corresponding waypoints)."

"The annotation of the flight plan is performed with exactly the same method as the one followed for raw aircraft trajectories."

"Our problem is informally described as the task of predicting the semantic trajectory of a flight between the departure and the destination airport, given a corresponding historic database of semantic trajectories and a flight plan."

The Proposed FSTP Framework

"Given the availability of both the flight plans F and the corresponding actual trajectories R, the emissions are designed as the deviation between these two, since this is the actual output observed from the evolution of a flight."

"The HMM approach for stage-2 in the proposed FSTP framework is based on an emissions model that translates deviations between flight plans and actual routes into probabilistic predictions in the maximum-likelihood sense."

"The continuous emissions model generates the 'most likely' per-waypoint deviation from the intended flight path F and, hence, the future route R. A more generic approach is to incorporate multiple dimensions in the input and also to approximate R itself, instead of the deviation from F. The simplest and straight-forward way to do this is by designing proper linear regressors (LR) [64] to replace the HMM counterparts for each waypoint."

Experiments and Results

"These design choices for the distance function were specifically selected as a compromise between clustering compactness versus ease of visualization, in order for the standard prediction error metrics MAPE and RMSE to be easily interpreted in the 3-D spatial-only sense."

"The per-waypoint prediction error remains fairly close to the mean (RMSE) value, not only in 3-D but also for each individual spatial dimension."

"This is particularly important, since these results prove the robustness of the NN-MLP predictions along the entire flight path and the validity of using the flight plan as the main element in constraint-based training."

"These numbers do not include stage-1 processing (clustering), which is in the order of several seconds, as this runs only once, regardless of the model selected for stage-2."

Discussion

"Combined with the partitioning of the input data via properly designed semantic-aware clustering, this modular approach is highly scalable and adaptable to any type of transit route and takeoff/landing patterns, provided that the associated flight plan is available."

"This is particularly important, since these results prove the robustness of the NN-MLP predictions along the entire flight path and the validity of using the flight plan as the main element of constraint-based training in the proposed FSTP framework."

"Even though deviations from the flight plan are common in mid-flight, especially when altering the landing approach due to changed weather conditions, the NN-MLP regressors model these deviations closely and follow the actual flight route, together with the trajectory grouping that clustering provides in the first stage of the proposed methodology."

"These three design aspects, i.e., independent per-waypoint model training and dimensionality reduction and input space partitioning via clustering, constitute this proposed approach inherently parallelizable and highly scalable to very large volumes and rates of data."

Conclusion

"This enrichment includes localized weather data (e.g., wind speed and direction), aircraft properties (e.g., type) and other external factors (e.g., weekday)."

"A set of independent predictive models are trained for each cluster, addressing the task of TP in the context of each reference waypoint of the flight plans."

"HMM, linear and nonlinear regressors were employed as base for the predictive models, exploring the trade-off between having very simple predictors and moderate accuracies versus more complex predictors and higher accuracies."

"The experimental results proved the feasibility of the proposed FSTP framework in real-world applications, in terms of prediction accuracy versus varying complexity (HMM, LR, NN)."

AHP-Entropy Based Priority Assessment of Factors to Reduce Aviation Fuel Consumption [65]

This is a machine-generated summary of:

Singh, Jagroop; Sharma, Somesh Kumar; Srivastava, Rajnish: AHP-Entropy based priority assessment of factors to reduce aviation fuel consumption [65]

Published in: International Journal of System Assurance Engineering and Management (2019)

Link to original: https://doi.org/10.1007/s13198-019-00758-0

If you want to cite the papers, please refer to the original.

For technical reasons we could not place the page where the original quote is coming from.

Abstract-Summary

"This study presents an integrated framework for aviation fuel consumption reduction, which will also limit its CO_2 emissions."

"Further, this research aims to categorize influential factors and examine their relative importance for fuel-efficient aviation."

"This study's theoretical framework combines and reconciles eight major areas: alternative jet fuels, aviation infrastructure, aircraft operations, socio-ecopolitical environment, aircraft design, technology, environmental uncertainty, and strategic changes."

"The priority ratings of these sub-factors with respect to 'aviation fuel consumption reduction' objective is measured by hybrid analytical hierarchy process-entropy method, using pair-wise comparison matrices."

"The findings attributed the highest importance to 'technological innovations', followed by 'aircraft design' and 'aircraft operations' for saving aviation fuel."

"This study shows the path for continuous improvement in aviation fuel efficiency by directing efforts and investments on highly important factors."

Introduction

"The way forward to reduce CO_2 emissions is by reducing aircraft fuel consumption or by improving overall aviation fuel efficiency."

"In the context of this study ratio of revenue passenger-kilometer (RPK) to fuel consumption is considered as a measure of aviation fuel efficiency."

"Lee and Mo [66] found that there is a significant correlation between aviation fuel consumption, operating cost of an airline, and environmental performance."

"In view of the benefits of reducing fuel consumption, numerous studies have acknowledged the role of facets in the aviation sector [67–74], yet the majority of studies remained focused on just a few factors."

"This research aims to achieve the following objectives: To investigate and categorize the aviation fuel consumption factors and sub-factors of the aviation industry; and To prioritize aviation fuel consumption factors based on their comparative importance."

Literature Review

"Various factors contribute towards improving aircraft fuel efficiency and alleviating environmental concerns by reducing CO_2 emissions."

"The aviation industry is scrutinizing every step of aircraft operations in the quest to cut its fuel cost."

"The rise in aviation fuel prices and taxation increases aircraft total operating cost, thus dictates the aviation industry to look and implement initiatives for fuel consumption reduction [74]."

"These socio-ecopolitical initiatives will boost the research, development, and use of better technologies and operations for fuel-efficient aviation."

"Same is the case with the aviation industry as technological innovations are credited for major aircraft fuel efficiency improvements."

"The aviation technology must progress continuously with an aim to improve aircraft fuel efficiency under prevailing growth and environmental scenarios."

"Pulling down airport operational efficiency and raising aircraft fuel consumption."

Research Methodology

"Pairwise comparisons are made in AHP to develop priorities amongst factors and sub-factors at respective hierarchical level."

"Following this AHP-Entropy method has been widely used for applications such as weight assessment [75], safety evaluation [76], risk assessment [77], sustainability assessment [78], and environment management [79]."

"Inspired by the widespread applications and the advantages offered by integrated AHP-Entropy, it has been applied in this study for prioritizing the fuel consumption factors in the aviation industry."

"AHP-Entropy integrated methodology has been used to investigate the comparative priorities of the factors and sub-factors from the aviation domain for achieving a reduction in its fuel burn."

"Phase-1 follows AHP procedure (Step A1–A5) to calculate AHP weights and check the consistency of factors."

"Phase-2 calculates entropy weights (Step B1–B3), which are integrated with AHP weights to obtain comprehensive weights (Step AB1)."

"To estimate an overall ranking for sub-factor, AHP-Entropy combines the priority weight of a factor with the comparison rating for respective sub-factors."

Results and Discussions

"Level-1 establishes the main objective of the study, i.e. 'Determination of influential factors for reduction in aviation fuel consumption'."

"Technological innovations' with priority percentage of 29.5, indicates that even now it is the main contributor towards aviation fuel consumption reduction."

"Third-ranked 'aircraft operations' with priority percent of 18.4 show that an effective aircraft design, with the incorporation of latest technological innovations, will not give best fuel efficiency and improved environmental performance until it is operated at best operational parameters [70]."

"With the total priority percentage of 81 'optimum flight profile' (0.347), 'optimum payload' (0.216), and 'air route availability' (0.172) emerged as main aircraft operational measures to achieve fuel consumption reduction."

"Based on the inferences from the results of this study, it can be stated that fuel saving is not only dependent on aircraft manufacturers but also requires involvement of all the actors, i.e. governments, airlines and operation managers."

Conclusions, Implications, and Limitations

"Among the numerous factors influencing aviation fuel consumption, some factors have direct effect whereas other factors either support or have an indirect effect."

"For the aviation industry, technology, strategies, infrastructure, aircraft design, and operations directly affect its fuel consumption; hence emerged as the top five influential factors."

"The aviation fuel consumption literature shows the scarceness of comprehensive research covering the overall domain of the aviation sector and priority factors for achieving aviation fuel saving."

"As it is difficult to invest equal efforts in each individual area of aviation due to limited resources, the findings provide insights regarding priority targets among factors and sub-factors to achieve aviation fuel savings."

"This study evolves the importance of the factors for aviation fuel consumption, thus establishing the foundation for drafting short-term goals and long-term goals based upon the possible savings and constraints (time and financial) associated with them."

Heart Disease Prediction Using Machine Learning Techniques [80]

This is a machine-generated summary of:

Shah, Devansh; Patel, Samir; Bharti, Santosh Kumar: Heart Disease Prediction using Machine Learning Techniques [80]

Published in: SN Computer Science (2020)
Link to original: https://doi.org/10.1007/s42979-020-00365-y

If you want to cite the papers, please refer to the original.

For technical reasons we could not place the page where the original quote is coming from.

Abstract-Summary

"Heart disease, alternatively known as cardiovascular disease, encases various conditions that impact the heart and is the primary basis of death worldwide over the span of the past few decades."

"It associates many risk factors in heart disease and a need of the time to get accurate, reliable, and sensible approaches to make an early diagnosis to achieve prompt management of the disease."

"Researchers apply several data mining and machine learning techniques to analyse huge complex medical data, helping healthcare professionals to predict heart disease."

"This research paper presents various attributes related to heart disease, and the model on basis of supervised learning algorithms as Naïve Bayes, decision tree, K-nearest neighbor, and random forest algorithm."

"It uses the existing dataset from the Cleveland database of UCI repository of heart disease patients."

"This research paper aims to envision the probability of developing heart disease in the patients."

Introduction

"Data mining refers to the extraction of required information from huge datasets in various fields such as the medical field, business field, and educational field."

"Data mining is exploring huge datasets to extract hidden crucial decision-making information from a collection of a past repository for future analysis."

"These data need mining by various machine learning algorithms."

"Medical data mining using classification algorithms provides clinical aid through analysis."

"It tests the classification algorithms to predict heart disease in patients [81]."

"Various data mining techniques such as regression, clustering, association rule and classification techniques like Naïve Bayes, decision tree, random forest and K-nearest neighbor are used to classify various heart disease attributes in predicting heart disease."

"The classification model is developed using classification algorithms for prediction of heart disease."

"A discussion of algorithms used for heart disease prediction, comparison among the existing systems is made."

Background

"This research paper uses classification techniques to predict heart disease."

"This section gives a portrayal of the related subjects like machine learning and its methods with brief descriptions, data pre-processing, evaluation measurements and description of the dataset used in this research."

"It trains machine learning algorithms using a training dataset to create a model."

"The model uses the new input data to predict heart disease."

"Using machine learning, it detects hidden patterns in the input dataset to build models."

"The model uses the new input data to predict heart disease and then tested for accuracy."

"Machine learning techniques are classified as: The model is trained on a dataset that is labelled."

"Training dataset trains our model while testing dataset functions as new data to get accuracy of the model."

"It does not use labelled dataset nor the results are associated with data, thus model learns from the experience."

Approach Methodology

"To accomplish the aim, we have discussed the use of various machine learning algorithms on the data set and dataset analysis is mentioned in this research paper."

"The data are then classified and split into training data set and test data set which is run on various algorithms to achieve accuracy score results."

"Naïve Bayes is a simple, easy to implement, and efficient classification algorithm that handles non-linear, complicated data."

"Decision tree is a classification algorithm that works on categorical as well as numerical data."

"This algorithm has higher accuracy in comparison to other algorithms as it analyzes the dataset in the tree-like graph."

"The data may be over classified and only one attribute is tested at a time for decision-making."

"Random forest algorithm has obtained an accuracy of 91.6% with Cleveland dataset in [82]."

Results and Analysis

"This research was done on supervised machine learning classification techniques using Naïve Bayes, decision tree, random forest, and K-nearest neighbor on UCI repository."

"Dataset was classified and split into a training set and a test set."

"Pre-processing of the data is done and supervised classification techniques such as Naïve Bayes, decision tree, K-nearest neighbor, and random forest are applied to get accuracy score."

"The accuracy score results of different classification techniques were noted using Python Programming for training and test data sets."

Conclusion

"The overall aim is to define various data mining techniques useful in effective heart disease prediction."

"I applied four data mining classification techniques, K-nearest neighbor, Naive Bayes, decision tree, and random forest."

"K-nearest neighbor, Naïve Bayes, and random forest are the algorithms showing the best results in this model."

"Considering the limitations of this study, there is a need to implement more complex and combination of models to get higher accuracy for early prediction of heart disease."

References

1. Kolley L, Rückert N, Kastner M, Jahn C, Fischer K (2022) Robust berth scheduling using machine learning for vessel arrival time prediction. Flex Serv Manuf J. https://doi.org/10.1007/s10696-022-09462-x
2. de León AD, Lalla-Ruiz E, Melián-Batista B, Moreno-Vega JM (2017) A machine learning-based system for berth scheduling at bulk terminals. Expert Syst Appl 87:170–182. https://doi.org/10.1016/j.eswa.2017.06.010
3. Cheimanoff N, Fontane F, Kitri MN, Tchernev N (2021) A reduced VNS based approach for the dynamic continuous berth allocation problem in bulk terminals with tidal constraints. Expert Syst Appl 168:114215. https://doi.org/10.1016/j.eswa.2020.114215
4. Parolas I (2016) ETA prediction for containerships at the port of Rotterdam using machine learning techniques. MSc Thesis, TU Delft.
5. Grida M, Lee C-Y (2018) An empirical model for estimating berth and sailing times of mega container ships. Marit Policy Manag 45:1078–1093. https://doi.org/10.1080/03088839.2018.1452304v
6. Liu C, Xiang X, Zheng L (2017) Two decision models for berth allocation problem under uncertainty considering service level. Flex Serv Manuf J 29:312–344. https://doi.org/10.1007/s10696-017-9295-5
7. Heilig L, Stahlbock R, Voß S (2020) From digitalization to data-driven decision making in container terminals. In: Böse JW (ed) Handbook of terminal planning, second edn. Springer, Cham, pp 125–154. https://doi.org/10.1007/978-3-030-39990-0_6
8. Giusti L, Carvalho L, Gomes AT, Coutinho R, Soares J, Ogasawara E (2022) Analyzing flight delay prediction under concept drift. Evol Syst. https://doi.org/10.1007/s12530-021-09415-z
9. Du WB, Zhang MY, Zhang Y, Cao XB, Zhang J (2018) Delay causality network in air transport systems. Transp Res Part E Logist Transp Rev. 118:466–476
10. Wu CL, Law K (2019) Modelling the delay propagation effects of multiple resource connections in an airline network using a Bayesian network model. Transp Res Part E Logist Transp Rev. 122:62–77
11. Kim Y, Choi S, Briceno S, Mavris D (2016) A deep learning approach to flight delay prediction. In: 2016 IEEE/AIAA 35th Digital avionics systems conference (DASC). Sacramento, CA, USA, pp. 1–6.
12. Gui G, Liu F, Sun J, Yang J, Zhou Z, Zhao D (2020) Flight delay prediction based on aviation big data and machine learning. IEEE Trans Veh Technol 69(1):140–150
13. Gama J, Zliobaite I, Bifet A, Pechenizkiy M, Bouchachia A (2014) A survey on concept drift adaptation. ACM Comput Surv 46(4):1–37

14. Webb G, Hyde R, Cao H, Nguyen H, Petitjean F (2016) Characterizing concept drift. Data Min Knowl Discov 30(4):964–994
15. Iwashita A, Papa J (2019) An overview on concept drift learning. IEEE Access 7:1532–1547
16. Lu J, Liu A, Dong F, Gu F, Gama J, Zhang G (2019) Learning under concept drift: a review. IEEE Trans Knowl Data Eng 31(12):2346–2363
17. Khamassi I, Sayed-Mouchaweh M (2014) Drift detection and monitoring in non-stationary environments. In: 2014 IEEE conference on evolving and adaptive intelligent systems (EAIS). Linz, Austria, pp. 1–6.
18. Pesaranghader A, Viktor H (2016) Fast hoeffding drift detection method for evolving data streams. In: 2016 Machine learning and knowledge discovery in databases (ECML-PKDD 2016). Riva del Garda, Italy, pp. 96–111.
19. Munoz Hernandez A, Scarlatti D, Costas P (2019) Real-time estimated time of arrival prediction system using historical surveillance data. In: 2019 45th Euromicro conference on software engineering and advanced applications (SEAA). Kallithea-Chalkidiki, Greece, pp. 174–177.
20. Wang K, Li J, Tian Y (2019) Airport delay prediction method based on improved weather impacted traffic index. In: Proceedings of 2019 IEEE first international conference on civil aviation safety and information technology. ICCASIT 2019, pp. 73–78.
21. Ai Y, Pan W, Yang C, Wu D, Tang J (2019) A deep learning approach to predict the spatial and temporal distribution of flight delay in network. J Intell Fuzzy Syst 37(5):6029–6037
22. Li MZ, Gopalakrishnan K, Balakrishnan H, Shin SH, Jalan D, Nandi A, Marla L (2021) Dynamics of disruption and recovery in air transportation networks. CEAS Aeronaut J. https://doi.org/10.1007/s13272-021-00521-x
23. Li MZ, Gopalakrishnan K, Pantoja K, Balakrishnan H (2021) Graph signal processing techniques for analyzing aviation disruptions. Trans. Sci. https://doi.org/10.1287/trsc.2020.1026
24. Ball M, Barnhart C, Nemhauser G, Odoni A (2007) Air transportation: Irregular operations and control. In: Handbooks in operations research and management science, vol 14, pp 1–67
25. Clausen J, Larsen A, Larsen J, Rezanova NJ (2010) Disruption management in the airline industry-concepts, models and methods. Comput. Oper. Res. 37(5):809–821
26. Li, M.Z., Gopalakrishnan, K., Wang, Y., Balakrishnan, H.: Outlier analysis of airport delay distributions in US and China. In: IEEE artificial intelligence and data analytics in air transportation conference (2020).
27. Yousefzadeh AM, Seyed Reza KT, Seyed Javad MC, Kheyrabadi M (2021) Optimization of air traffic management efficiency based on deep learning enriched by the long short-term memory (LSTM) and extreme learning machine (ELM). J Big Data. https://doi.org/10.1186/s40537-021-00438-6
28. Kim YJ (2017) A deep learning and parallel simulation methodology for air traffic management. Georgia Institute of Technology, Atlanta
29. Polvara R, et al. Toward end-to-end control for UAV autonomous landing via deep reinforcement learning. In: 2018 International conference on unmanned aircraft systems (ICUAS), IEEE. 2018.
30. Ni X et al (2019) Civil aviation safety evaluation based on deep belief network and principal component analysis. Saf Sci. 112:90–95
31. Yang H-F et al (2017) Optimized structure of the traffic flow forecasting model with a deep learning approach. IEEE Trans Neural Netw Learn Syst. 28(10):2371–2381
32. Tong C et al (2018) An innovative deep architecture for aircraft hard landing prediction based on time-series sensor data. Appl Soft Comput. 73:344–349
33. Venkatesh V, et al. Iterative machine and deep learning approach for aviation delay prediction. 978-1-5386-3004-4/17/$31.00 ©2017 IEEE. 2017.
34. Choi S (2019) A multi-level predictive methodology for terminal area air traffic flow. Georgia Institute of Technology
35. Kim YJ, Choi S, et al. A deep learning approach to flight delay prediction. 978–1–5090-2523-7/16/$31.00 ©2016 IEEE. 2016.

36. Tong C et al (2018) A novel deep learning method for aircraft landing speed prediction based on cloud-based sensor data. Future Gen Comput Syst. 88:552–558
37. Naessens H, et al. Predicting flight routes with a Deep Neural Network in the operational Air Traffic Flow and Capacity Management system. 17 December 2017 EUROCONTROL Maastricht Upper Area Control Centre Horsterweg 11, NL-6199 AC Maastricht Airport. 2019.
38. Lin Y et al (2019) Deep learning based short-term air traffic flow prediction considering temporal–spatial correlation. Aerosp Sci Technol. m5G; v1.256; Prn:17/04/2019; 12:58] P.1(1–9)
39. Zhang H, et al. Airport delay prediction based on spatiotemporal analysis and bi-LSTM sequence learning. 978-1-7281-4094-0/19/$31.00 ©2019 IEEE. 2019.
40. Yazdi MF, Kamel SR, Chabok SJM, Kheirabadi M (2020) Flight delay prediction based on deep learning and Levenberg-Marquart algorithm. J Big Data. https://doi.org/10.1186/s40537-020-00380-z
41. Choi S, et al. Prediction of weather-induced airline delays based on machine learning algorithms. In 2016 IEEE/AIAA 35th Digital Avionics Systems Conference (DASC). New York: IEEE. 2016
42. Rebollo J, Balakrishnan H (2014) Characterization and prediction of air traffic delays. Transp Res Part C Emerg Technol 44:231–241
43. Oza S et al (2015) Flight delay prediction system using weighted multiple linear regression. Int J Eng Comp Sci. 4(05):11765
44. Thiagarajan B, et al. A machine learning approach for prediction of on-time performance of flights. In 2017 IEEE/AIAA 36th Digital Avionics Systems Conference (DASC). New York: IEEE. 2017.
45. Al-Tabbakh SM, El-Zahed H (2018) Machine learning techniques for analysis of Egyptian Flight delay. J Sci Res Sci. 35(1):390–399
46. Bishop CM (2016) Pattern recognition and machine learning. Springer, Berlin
47. Kim YJ, et al. A deep learning approach to flight delay prediction. In 2016 IEEE/AIAA 35th Digital Avionics Systems Conference (DASC). New York: IEEE. 2016.
48. Venkatesh V, et al. Iterative machine and deep learning approach for aviation delay prediction. In 2017 feourth IEEE Uttar Pradesh Section International Conference on Electrical, Computer and Electronics (UPCON). New York: IEEE. 2017.
49. Meel P, et al. Predicting flight delays with error calculation using machine learned classifiers. In 2020 seventh International Conference on Signal Processing and Integrated Networks (SPIN). New York: IEEE. 2020.
50. Yu B, Guo Z, Asian S, Wang H, Chen G (2019) Flight delay prediction for commercial air transport: a deep learning approach. Transp Res Part E Logist Transp Rev. 125:203–221
51. Yu H, Wilamowski BM (2011) Levenberg-Marquardt training. In: Industrial electronics handbook. Routledge, London, p 1
52. Chawla NV (2009) Data mining for imbalanced datasets: an overview, In Data mining and knowledge discovery handbook. Springer, Berlin, pp 875–886
53. Qu J, Zhao T, Ye M, Li J, Liu C (2020) Flight delay prediction using deep convolutional neural network based on fusion of meteorological data. Neural Process Lett. https://doi.org/10.1007/s11063-020-10,318-4
54. Cao WD, Lin XY (2011) Forecast warning level of flight delays based on incremental ranking support vector machine. Comput Eng Des 32(5):1770–1776
55. Khanmohammadi S, Tutun S, Kucuk Y (2016) A new multilevel input layer artificial neural network for predicting flight delays at JFK airport. Complex adaptive systems, Los Angeles, USA, pp. 237–244.
56. Loffe S, Szegedy C (2015) Batch normalization: accelerating deep network training by reducing internal covariate shift. In: ICML 2015. Lile, France, pp 448–456
57. Aljubairy A, Zhang WE, Shemshadi A, Mahmood A, Sheng QZ (2020) A system for effectively predicting flight delays based on IoT data. Computing. https://doi.org/10.1007/s00607-020-00794-w

58. Aljubairy A, Shemshadi A, Sheng QZ (2016) Real-time investigation of flight delays based on the Internet of Things data. In: Proceedings of international conference on advanced data mining and applications (ADMA), pp. 788–800.
59. Rosenberger JM, Schaefer AJ, Goldsman D, Johnson EL, Kleywegt AJ, Nemhauser GL (2002) A stochastic model of airline operations. Transp Sci 36(4):357–377
60. Horiguchi Y, Baba Y, Kashima H, Suzuki M, Kayahara H, Maeno J (2017) Predicting fuel consumption and flight delays for low-cost airlines. In: Proceedings of the 31st AAAI conference on artificial intelligence (AAAI), pp. 4686–4693.
61. Ayhan S, Costas P, Samet H (2018) Predicting estimated time of arrival for commercial flights. In: Proceedings of the 24th ACM SIGKDD international conference on knowledge discovery and data mining, pp. 33–42.
62. Georgiou H, Pelekis N, Sideridis S, Scarlatti D, Theodoridis Y (2019) Semantic-aware aircraft trajectory prediction using flight plans. Int J Data Sci Anal. https://doi.org/10.1007/s41060-019-00182-4
63. Ayhan, S., Samet, H.: Aircraft trajectory prediction made easy with predictive analytics. In: Proceedings of ACM SIGKDD 2016 (2016).
64. Theodoridis S, Koutroumbas K (2008) Pattern Recognition, fourth edn. Academic Press, Cambridge
65. Singh J, Sharma SK, Srivastava R (2019) AHP-Entropy based priority assessment of factors to reduce aviation fuel consumption. Int J Syst Assur Eng Manag. https://doi.org/10.1007/s13198-019-00758-0
66. Lee J, Mo J (2011) Analysis of technological innovation and environmental performance improvement in aviation sector. Int J Environ Res Public Health 8(9):3777–3795. https://doi.org/10.3390/ijerph8093777
67. Brueckner J, Abreu C (2017) Airline fuel usage and carbon emissions: determining factors. J Air Transp Manag 62:10–17. https://doi.org/10.1016/j.jairtraman.2017.01.004
68. Cansino J, Román R (2017) Energy efficiency improvements in air traffic: the case of airbus A320 in Spain. Energy Policy 101:109–122. https://doi.org/10.1016/j.enpol.2016.11.027
69. Drake J (1974) Social, political and economic constraints on airline fuel optimization. Transp Res 8(4–5):443–449. https://doi.org/10.1016/0041-1647(74)90064-1
70. Grote M, Williams I, Preston J (2014) Direct carbon dioxide emissions from civil aircraft. Atmos Environ 95:214–224. https://doi.org/10.1016/j.atmosenv.2014.06.042
71. Lee J (2010) Can we accelerate the improvement of energy efficiency in aircraft systems? Energy Convers Manag 51(1):189–196. https://doi.org/10.1016/j.enconman.2009.09.011
72. Müller C, Kieckhäfer K, Spengler T (2018) The influence of emission thresholds and retrofit options on airline fleet planning: an optimization approach. Energy Policy 112:242–257. https://doi.org/10.1016/j.enpol.2017.10.022
73. Reynolds TG (2015) Green aircraft operations. Encycl Aerosp Eng:1–12
74. Sarkar A (2012) Evolving green aviation transport system: a holistic approach to sustainable green market development. AJCC 01(03):164–180. https://doi.org/10.4236/ajcc.2012.13014
75. Al-Aomar R (2010) A combined ahp-entropy method for deriving subjective and objective criteria weights. Int J Ind Eng: Theory Appl Pract 17(1):12–24
76. Chuansheng X, Dapeng D, Shengping H, Xin X, Yingjie C (2012) Safety evaluation of smart grid based on AHP-entropy method. Syst Eng Procedia 4:203–209. https://doi.org/10.1016/j.sepro.2011.11.067
77. Wang T (2014) The research of electronic banking risk evaluation based on comprehensive assessment AHP-Entropy. Int J u- e-Serv Sci Technol 7(6):413–422. https://doi.org/10.14257/ijunesst.2014.7.6.36
78. Wu G, Duan K, Zuo J, Zhao X, Tang D (2017) Integrated sustainability assessment of public rental housing community based on a hybrid method of AHP-Entropy weight and cloud model. Sustainability 9(4):603. https://doi.org/10.3390/su9040603

79. Bai L, Wang H, Huang N, Du Q, Huang Y (2018) An environmental management maturity model of construction programs using the AHP-entropy approach. Int J Environ Res Public Health 15(7):1317. https://doi.org/10.3390/ijerph15071317
80. Shah D, Patel S, Bharti SK (2020) Heart disease prediction using machine learning techniques. SN Computer Science. https://doi.org/10.1007/s42979-020-00365-y
81. Ramalingam VV, Dandapath A, Raja MK (2018) Heart disease prediction using machine learning techniques: a survey. Int J Eng Technol. 7(2.8):684–687
82. Xu S, Zhang Z, Wang D, Hu J, Duan X, Zhu T. Cardiovascular risk prediction method based on CFS subset evaluation and random forest classification framework. In: 2017 IEEE second international conference on big data analysis (ICBDA). IEEE. p. 228–232.

Conclusion by the Editor

After reading all the chapters from this book, it has taught readers to learn the depth of knowledge and wisdom behind CAMS. This book ensures that the learners get a feeling of how to design, develop, prototype, and commercialize the products or solutions towards CAMS.

Chapter 1 provides an overview of CAMS, which helps learners understand AI models with the background of reinforcement learning and AI-based 2D and 3D models. By and large, most of the learners have design understanding, analogical reasoning, and knowledge transfer from this book chapter.

Chapter 2 addresses the computations and methods that enable learners to solve AI-related problems in various environments. After digesting this chapter, the readers will have a dynamic sensing approach to solve clustering problems for robots or drones. Subsequently, they have a wide knowledge on multiagent reinforcement learning with applications.

Chapter 3 addresses the industrial challenges and solutions that promote learners to imbibe cutting-edge industrial techno activities. After learning this chapter, all readers could have a research aptitude to tackle challenges such as medical image denoising, wavelet multiscale analysis, EEG signals to digit classification, low-energy FFT architectures, deep learning classification of motherboard components and optimization of VMD.

Chapter 4 addresses the future trends of CAMS, which envisions learners to scale up new applications in multiple domains. After analyzing this chapter, the readers have an innovative mindset to go for new products or solutions as per industry needs. From this chapter, the readers will come to know how to improvise non-deterministic polynomial problems described with solutions through machine learning methods.

At the end of all the chapters, the readers will have a wide scope on CAMS with their applications.

V. Natarajan (ed.), *Computational Artificial Intelligence and Methods for industries*, https://doi.org/10.1007/978-981-96-5277-8

In the next version of the book, the author would like to bring out the conversational AI models through CAMS such as ChatGPT, Google Gemini (formerly Bard), Jasper AI, DeepSeek and Bing chat. Happy to receive the feedback of this book through the publisher. Hope for the best!

The manufacturer's authorised representative in the EU is Springer Nature Customer Service Centre GmbH, Europaplatz 3, 69115 Heidelberg, Germany. If you have any concerns regarding our products, please contact ProductSafety@springernature.com

Printed and bound by CPI Group (UK) Ltd, Croydon, CR0 4YY
01/10/2026
02225850-0011